FISH FEEL PAIN!

To Regina R.I.P.

FISH FEEL
PAIN!
Scrutiny of a dogma

Alexander Schwab

MERLIN UNWIN
BOOKS

Merlin Unwin Books Ltd
Palmers House
7 Corve Street
Ludlow
Shropshire SY8 1DB
UK

www.merlinunwin.co.uk

ISBN 978 1 913159 73 3
Typeset in 11.5 point Minion Pro by Merlin Unwin Books
Printed by Bell & Bain Ltd, UK

Contents

Introduction

Fish feel pain (FFP) is a persuasive dogma declared by the progressive orthodoxy of animal rights and political correctness. It derives its plausibility by reference to science. Worldwide the FFP dogma fuels the demand not just for restrictions and bans on recreational fishing but on all fish use including commercial fishing, aquaculture and aquaristics.

The dogma is: 'fish feel pain' – end of story. It's on thousands of websites, articles and reports. On the basis of the FFP dogma, welfare legislations in jurisdictions the world over classify fish as 'sentient' i.e. capable of feeling pain. In New Zealand and the UK, for example, fish, squid, crabs and lobsters are considered sentient. Animal rights advocates point out that if fish sentience were to be taken seriously – as enshrined in animal welfare laws – that would be the end of fish use altogether. While that extreme seems unrealistic currently, recreational fishing is a soft target for piecemeal restrictions and bans. That is exactly what is happening – repeatedly – the world over.

The dogma emerged shortly after the publication of a ground-breaking study by Lynne U. Sneddon in 2003 which purported to show that fish feel pain. In the same year Merlin Unwin Books, Ludlow, England, published my book *Hook, Line and Thinker – Angling and Ethics*. This was a pioneering achievement: up to that moment nobody had attempted to address the history and philosophy of the ethics of angling in a systematic way. It was also intended as a wake-up call to the burgeoning power of animal rights. In vain. Now 20 years later the FFP science/animal rights theorists hold the baton and dictate the tune.

More than that, the movement has actually bred a new discipline and a new species: animal law theory and animal lawyers. Entire university departments are dedicated to 'animal law'. The crucial development in this is that animal law minds melt down contradictory philosophical theories and fuse them into legal language peppered with appeals to compassion. The radical and misanthropic philosophical roots are obscured: it's all smoke and mirrors.

A quick look back

For the world's estimated 200 million anglers, Wednesday 30 April 2003 was a day like no other. On that date The Royal Society published an article by Lynne Sneddon with the innocuous title '*Do fishes have nociceptors? Evidence for the evolution of a vertebrate sensory system*'. It immediately sent shockwaves through the world of fishing and beyond and continues to do so. The important message of that article was that fish feel pain. The world's media quickly latched on to the logical consequences. As *The Independent* newspaper in the UK put it succinctly: 'Anglers on the hook as study says fish feel pain'.

The topic had been in the air for some time. Unbeknown to each other, two scientists and one angler had been working on the same cluster of issues relating to fish pain. In early 2002 scientist James Rose addressed the question of fish pain and concluded that fish do not feel pain. Rose's article was published shortly before my book *Hook, Line and Thinker – Angling and Ethics* went to press and it confirmed all my arguments on the subject. But it wasn't until the following year, 2003, that Lynne Sneddon's seminal nociceptor article catapulted the issue into the headlines. The cruelty charge had been floating around for a while but Sneddon's findings gave and continue to give the charge a completely different quality and urgency.

It's not that the FFP-question is new. Lord Byron's (1788–1824) lines still echo in today's anti-angling propaganda. Lord Byron's go at Izaak Walton has lost none of its punch. Like it or not, it's very witty.

And angling, too, that solitary vice,
Whatever Izaak Walton sings or says:
The quaint, old, cruel coxcomb, in his gullet
Should have a hook, and a small trout to pull it.

The Black Pennell must be one of the most popular flies all over the world. However few anglers will be familiar with Cholmondeley-Pennell's work on fish pain. Henry Cholmondeley-Pennell (1837–1915) addressed the FFP issue not with poetry but with science. His observations, thoughts and conclusions are as relevant today as they were then and deserve to be resurrected from obscurity.

The contemporary possibilities of science activists and the high-flying philosophical exhortations of animal rights philosophers were unheard of in Henry Cholmondeley-Pennell's day. My meanderings in this book are both an update and a footnote on Cholmondeley-Pennell's thoughts on the matter.

At the end of this introduction there is a reprint of the original 1870 text. In Chapter 8 you'll also find the most up-to-date scientific take on the FFP-issue. While Cholmondeley-Pennell's contribution is couched in plain language, the 'Reasons to be skeptical about sentience and pain in fishes and aquatic invertebrates' is penned by a group of leading scientists from across the world. If you want to follow up the scientific angle of the FFP-issue, this is a very useful source of information.

[Price Sixpence.

CAN FISH FEEL PAIN?

THE QUESTION CONSIDERED ANALOGICALLY AND
PHYSIOLOGICALLY;

WITH

A FEW WORDS ON THE ETHICS OF

ANGLING.

BY

H. CHOLMONDELEY-PENNELL,

AUTHOR OF "THE MODERN PRACTICAL ANGLER,"
"THE ANGLER-NATURALIST," "THE BOOK OF THE PIKE," &c.

LONDON:
FREDERICK WARNE AND CO.
BEDFORD STREET, COVENT GARDEN.
1870.

In the 20th century you find here and there some ethical snippets in angling literature, most notably in A.A. Luce's classic *Fishing and Thinking* (1959) and in Bryn Hammond's *Halcyon Days* (1992). Apart from those markers, the ethics of recreational angling were largely uncharted territory. Pertinent and timely as *Hook, Line and Thinker* may have been, only an infinitesimal small number of anglers showed interest because the subject matter didn't really add to the pleasure of fishing in any way. It isn't unduly exaggerated to say that *Hook, Line and*

Thinker fell practically unnoticed from the press. Worse than that was its defiant optimism which held that once people and lawmakers understood the fundamental misanthropy of animal rights the tide would be turning. Far from it: in the 20 years since the publication of *Hook, Line and Thinker* the animal rights movement worldwide has gone from strength to strength. Animal rights lingo and concepts have made it, for example, into the Treaty of Lisbon (European constitution) and one of its spin-offs made it into UK law (2022) recognising crabs, lobsters and octopuses as 'sentient' beings.

According to a 2021 survey, recreational fishing in First World countries seems to be still widely tolerated. Nevertheless the demands for a ban either on certain practices (e.g. catch-and-release, live bait, keep nets) or on the whole activity continue. In fact Germany and Switzerland have implemented laws based on the FFP-assumption. Given a favourable political conjunction or some legal absurdity, it's not entirely unfeasible to see recreational fishing banned in certain countries or regions. It happened in Colombia where the high court ruled that sportfishing is unconstitutional. The reasoning of the court:

> Although there is no consensus as to whether fish are sentient beings, the truth is that by virtue of the precautionary principle, even in the absence of scientific certainty [...] the intervention of the State is necessary.

Anti-angling won't just go away. Too many influential people and organisations are committed to the cause and committed to creating an atmosphere that is hostile to fishing – not least to discourage people from going fishing.

Why didn't I spare you this?

Why on earth should I punish myself and you, dear reader, by writing about fish pain again and by revisiting *Hook, Line and Thinker* twenty years after its publication? Indeed, why do I call it punishment? The reason is straightforward: as most of the themes touch on philosophy and specifically animal rights philosophy there isn't much fun to be had. Unless you are reading this in the close season it's a waste of valuable fishing time. Worse than that, some of the stuff in contemporary philosophy is unbearably depressing. As I wrote in *Hook, Line and Thinker* in 2003: 'There's a morbidness oozing out of some philosophy books that is positively revolting.'

In 2003 I observed:

> The last twenty years have witnessed an almost unchecked international proliferation of animal rights ideas. Many animal welfare societies and environmental organisations have drifted slowly but steadily from the concept of welfare to rights. Animal rights have also gained a foothold in the educational and political establishment.

Twenty years later they *are* the educational and political establishment. 'They' meaning the activist elites of major universities and political parties and their agendas, beliefs and dogmas.

Animal rights has trickled from top down into all layers of Western societies. As the supermarket shelves everywhere in the First World testify, the affluence-driven, ethical diet has established itself and it keeps growing. Vegetarianism and veganism are *de rigueur* for the virtue-intoxicated animal rights missionaries.

If you can't beat them, join them? Not if I can help it because the fundamental misanthropy, pessimism and authoritarianism by which the animal rights movement is driven is too much to swallow.

So burying my head in the sand is not on the cards and there might be likewise a couple of fellow anglers and hunters who are interested in what the score is today. Other intrepid readers might be a new generation of editors of fishing and other outdoor magazines, websites, blogs or internet channels, fisheries managers and nature/wildlife protection professionals of all kinds.

These probably are the people most often confronted with practical animal rights issues and my take on animal rights might be in a small way useful to them.

The three main points contended in *Hook, Line and Thinker* were:

1. Anti-recreational fishing and all anti-fishing is based on animal rights philosophy which in turn is embedded in a more general movement for a better, cruelty-free, just, all-inclusive LGBTQQIP2SAA+UY world. *Hook, Line and Thinker*: 'Fundamental societal change is indeed what the leading animal rights philosophers and their zealots have in mind.'

2. The presumed cruel angler stands in the way of a better world. The moral sell-by date of anglers (and all other animal users) has expired and their place is in the historical trash bin. *Hook, Line and Thinker*: 'Cruelty in angling is merely a means to an end. Anti-anglers are not interested in fish...'

3. Fish pain is pivotal to the anti-fishing case. Cruelty is a deliberate act causing pain and suffering. *Hook, Line and Thinker*: 'No pain, no suffering, no cruelty.' Without suffering, the entire animal rights case collapses.

Such is still the case. However, the scientific FFP claim as put forward in Lynne Sneddon's article took the FFP question to the next level.

Philosophers, social scientists and activists have since jumped on the bandwagon and it's not just presumed angling cruelty in the pillory but all fish use including commercial and subsistence fishing, aquaculture, aquaristics and science.

Now in 2023 I see that the same themes have evolved and gained in contour:

1. Animal rights is absolutely hostile to recreational fishing. It is a pessimistic, misanthropic and life-denying message. It's monochromatic and full of ugliness.

2. Recreational fishing is part of conservation which is an optimistic, philanthropic and life-affirming message. It's polychromatic and full of beauty. Indeed, going fishing is the quest for beauty.

3. The new kid on the block is the FFP scientist activist teaming up with philosophers, lawyers and politicians for animal rights agitprop.

4. Legal theory has become the spearhead of the animal rights movement because it can easily fuse and accommodate mutually exclusive philosophical concepts. It creates a comfort zone for politicians and prepares the ground for legislation.

For this book and update I borrow freely and paraphrase from my previous publications since 2003. I have omitted all earlier errors and shortcomings and if I have replaced them with new ones I exclusively take the blame. Like *Hook, Line and Thinker* in 2003 this is rough and ready stuff. If you require detailed source information please visit www.philosofish.ch where you can contact me.

Why are we here, twenty years later, still embroiled in this agonising debate about whether fish feel pain? Animal rights philosophy doesn't evolve, it spreads with substantially the same arguments over and over and over again. And there is no end in sight, no sell-by date. It's pure purgatory – perhaps a 'lite' purgatory because it has novelty value if you're unfamiliar with the subject.

Meanwhile spare a thought for the poor lab fish which supposedly feel pain yet are still experimented on after all this time. It is exactly those scientists who believe (they don't *know*) that fish feel pain who happily keep experimenting on fish, only to come time and again to their same conclusion: fish feel pain. But they knew that before, so why keep on torturing the defenceless creatures? It all runs in a very sophisticated jargon-heavy loop with no end in sight. But content is one thing. Politics is another. As far as politics is concerned, animal rights and FFP science are very real and very influential.

Not long after the publication of Lynne Sneddon's article, some scientists teamed up with philosophers. The result was that an open scientific question morphed into a predominantly philosophical, ethical and political issue. 'Ethical' because of the potential cruelty involved in all fish use and 'philosophical' because of the consciousness angle. Ethics and the philosophy of mind are a free-for-all territory. Moralising and wild speculation in support of the FFP claim know no boundaries. The reason the FFP claim is so popular with philosophers, activists, politicians and journalists is because it can be conveniently embedded in any kind of agenda.

These days it is frequently asserted that there is a 'consensus' among scientists that fish feel pain and that FFP is thus a fact. 'Consensus' means that there is no conclusive evidence, otherwise there would be no need for 'consensus'. If you try hard enough, you can find a consensus group for anything. Scientific, philosophical and legal consensus can declare any nonsense as fact. The current scientific, philosophical and legal group 'consensus' for the generic claim that fish feel pain is like saying that the earth is pretzel-shaped. Scepticism about fish feel pain consensus can easily be brushed off.

As one philosopher has it:

The arguments of the biologists against fish pain are philosophically not conclusive.

The arguments of astronomers against the pretzel-shaped planet are philosophically not conclusive.

Ideology trumps science. There is definitely a whiff of *1984* about all this.

Professors James D. Rose and Brian Key have tried to keep me away from too many factual errors and I could also draw on Dr. Ben Diggles' in-depth knowledge of the issue. They have done their best to keep me out of trouble. If they haven't succeeded it's entirely of my doing. Acknowledging and thanking them here does not mean they in any way endorse my interpretations or views.

Finally: this is a piece of angling literature, not a philosophical or scientific contribution in the narrow sense. I am therefore at liberty to go off on a tangent at times, in the hope that wisdom might be found somewhere along the line.

Alex Schwab, Nelson, New Zealand, 30 August 2023

Henry Cholmondeley-Pennell (1870)
The Ethics of Angling:
CAN FISH FEEL PAIN?

THE charge of 'cruelty' as brought against angling has been so often and so ably answered, that it would almost seem to be unworthy of serious refutation; but as it has lately been revived by certain writers in the press, whereby the 'consciences of weak brethren' are vexed, a few plain words on the moral view of the subject, as well as on its physiological aspect, may not be an inappropriate companion to my general work on fishing*, which is published to-day.

Of fish, in common with 'every moving thing that liveth,' it has been said to man, 'Into your hand they are delivered . . . they shall be meat for you.' Without propounding a paradox, Death is the law of their Life: and whether it is inflicted by the net or the rod, whether the notion of 'sport' be or be not associated with such infliction, it makes no difference in the essence of the act itself. This law has been universally recognised in all ages, by all people, amongst whom 'mighty hunters' have received honours secondary only to those of mighty warriors—the two qualifications being, indeed, constantly united. It was necessary to hunt to live, and doubtless the instinct of sport was wisely given as incentive to an occupation in many cases essential to existence. The same instinct is observed unmistakably amongst the inferior orders of predatory animals themselves.

---

* *The Modern Practical Angler:* A Complete Guide to Flyfishing Trolling, and Bottom-fishing. Illustrated by Fifty Engravings of Fish and Tackle. Cloth, 6s. London: Frederick Warne and Co. Bedford Street, Covent Garden.

There is, therefore, *ipse facto*, no cruelty in inflicting death on fish or any other animals, provided unnecessary pain or intentional waste is avoided. I say 'waste,' because I hold, in common with the best sportsmen of modern times, that to destroy life (except noxious life) wantonly, that is, without prospect of the destruction subserving any useful purpose, would clearly be an abuse of the authority vested in man—in other words, would be an act of cruelty.

But to the charge of such wanton destruction, anglers are clearly not open; for as a rule all fish caught by the rod are eaten, and thus accomplish what was doubtless one of the primary objects of their creation. So then, if cruelty ever takes place, the responsibility of it is chargeable not upon angling, *quoad* angling, but upon individual fishermen violating its canons; and proves no more than the malversation of any other lawful pursuit would prove against those who carry it on legitimately. It is not the use but the abuse of the art.

It follows from these observations, if correct, that even if the infliction of physical suffering were an unavoidable incident to the catching and killing of fish, fishing would still be perfectly legitimate. It is my firm belief, however, that in point of fact such suffering is not so inflicted; I believe that pain, in the sense in which human beings are conscious of it, is unknown to fish organization; and to show what reasons there are for such belief is my principal object in this paper.

These reasons are based upon analogy and physiology; upon one or other of which bases all the arguments of those who maintain the opposite theory—that fish do suffer pain—must necessarily rest.

Their case would probably be expressed thus:—'We know, as a physiological fact, that we ourselves—human beings—suffer pain, producing in most cases certain overt phenomena, strugglings, contortions, and so forth. We see similar phenomena in the case of fish hooked, maimed, or taken out of water, and we argue by analogy that they are the result of a similar cause. Moreover, we know that fish do possess nerves of sensation, and, therefore, we believe that their organization is capable of experiencing pain.'

More frequently the whole question is begged, and the denouncers of the cruelty of angling start by assuming as a postulate the position which they ought to demonstrate as a premise. When, however, any proof is attempted to be adduced, the above, varied, perhaps, in form, will be found to be the sum and substance of it. If, therefore, these premises are fallacious, the conclusion based upon them fails also. In other words, if it can be shown that the overt phenomena referred to, have not, of necessity, any connexion whatever with pain, there will be no *evidence* of the existence of pain itself; and if it can be further shown that all the deductions, both of physiology and analogy, point exactly the other way, there will, according to their own chain of reasoning, be sufficient presumption to justify an opposite inference.

I will take the last point in their argument—the supposed connexion between pain and sensation—first. Now physiology clearly shows that, as a matter of fact, no such connexion of necessity exists. Sensation is merely that power by which fish, in common with all other organisms, are enabled to receive impressions from external objects; and although to intensify sensation beyond a certain point, might in the human subject amount to pain, pain itself does not consist in the mere exaggeration of ordinary sensitive impressions, but is of a distinct and, as it were, superadded character, so that the capacity for the one may exist in a very acute degree without the capacity for the other. This is proved in the case of the administration of chloroform, which may be so given that the patient shall remain perfectly conscious, and even capable of feeling an operation performed, whilst suffering no pain whatsoever. Therefore, admitting, for the sake of argument, that the analogy holds good between a human being and a fish, the possession by the latter of the nerves of sensation, touch, taste, &c., in even the highest degree of development, does not of necessity imply any corresponding capacity of suffering; and this argument—the presumed connexion between pain and sensation—may accordingly be eliminated from the discussion.

The second point—the argument on which the asserters of fish-suffering mainly rest—is, as I have said, the overt phenomena or symptoms of pain,—the struggles, writhings, contortions, &c., observable when fish are hooked, injured, or taken out of water.

Now, one of the first axioms physiology teaches is that these phenomena are simply the effect of what is termed in scientific phraseology *a reflex action of the nerves*—the effect, that is, of a certain action transmitted by the nerves to the spinal marrow, occasioning violent movement, but which may be entirely painless, or even involuntary. A man being hung, or a patient in an epileptic fit, might be cited as examples of the action of this law. In both cases violent muscular contortions and even convulsions occur, arguing to the natural mind a corresponding intensity of suffering, and yet, viewed in the light of physiology, there is every reason for believing that they are wholly unaccompanied by pain. The appearance witnessed in a case of St. Vitus' dance, where the most wayward play of the muscles takes place, is another familiar illustration.

As a matter of fact, fish strugglings and contortions are easily accounted for, without the supposition of pain, on grounds in analogy with every-day experience of other phases of animal life. Thus, for example, the natural instinct of any wild creature becoming suddenly conscious of restraint is at once to exert its utmost strength to escape,—witness the efforts of the wild horse under the lasso, or the struggles of the noosed elephant; and yet in neither of these cases is any actual pain inflicted. Even with semi-domesticated animals the same instinct may be observed, of which the struggles of a sheep caught for shearing or washing will be a familiar illustration. If, then, animals so much more elevated in the scale of intelligence struggle in this manner when restrained—although the restraint be totally unaccompanied by pain—why should not fish do the same? Terror is another obvious explanation.

The gasping of a fish out of water is also readily accounted for without any necessary pre-supposition of actual pain. The fish breathes by his gills; absence of moisture makes these stick together, and the gasping is merely the effort of nature to separate them. When the structure of the gills enables the fish to retain sufficient moisture for the purpose, no such gaspings ensue; and in the Broads of Norfolk, where tench are taken by tickling with the hand, these fish may often be lifted out of the water perfectly quiescent, and will so remain until roughly moved or disturbed.

It has been shown, therefore, (1) that sensation is not *necessarily* accompanied by a capacity for feeling pain, and (2) that strugglings, contortions, &c., are *of themselves* no evidence of its presence.

So far the negative argument. The next step is to examine what grounds there are for arriving at positive conclusions on these points. And here, again, physiology supplies us with two most important facts. The first fact is that the perception of pain—another word for pain itself—is entirely a cerebral, or brain function; and that whereas in the human organism the volume of brain as compared to the rest of the body is as 1 to 51, in a fish it is sometimes as low as 1 in

3700, or 74 times less: the second fact, perhaps even more significant, is that the blood of a human being is warm, whilst that of a fish is cold—colder in almost every case than the surrounding air; the blood globules also differing signally in shape. What effect these latter differences of constitution may have upon the respective capacities of feeling pain can only be conjectured: clearly opposed to the idea of any high degree of nervous sensibility. Therefore as far as physiology goes it is all in favour of the capacities of fish for suffering, if existing at all, existing only in the lowest and most rudimentary degree—in a degree so low, in fact, as almost to amount to a change in the essence of the thing itself. I do not wish, however, in any way to overstrain the physiological argument; nor, indeed, is there any necessity for doing so, as whatever links in the chain physiology leaves blank, analogy more than supplies.

Let us, then, in considering the analogical side of the argument, take human capacity of pain or suffering as the positive standard—as that, I mean, which we can apprehend—and let us examine what pain, or rather what phenomena of pain, follow what injuries: for although, as has been seen, overt phenomena are, of themselves, no proofs of pain, pain is almost always accompanied by such manifestations. Let us further see whether injuries to the inferior creature are followed by symptoms in any way corresponding to those which would be observed in the superior. If it is found that they are not;—if the external symptoms, the almost invariable accompaniments of physical pain, are not present—we may fairly reason by analogy that the pain itself is not present.

Of all such overt symptoms, appetite—the desire of eating—will probably be the most convenient to select, as it is an act generally indicative in all classes of animals of a certain amount of bodily ease and well being. The Pike, which is both a very voracious fish, and one occupying a high position in the scale of fish intelligence, as gauged by the relative volume of brain, will furnish an apt subject for a first comparison. What would be the effect upon the human subject, of, say, a couple of meat hooks being firmly imbedded in his jaws? Certainly the last thing he would feel any inclination to do, under such circumstances, would be to eat. And yet what is the effect which such an appendage has upon the Pike? The experience of most trollers will provide a reply, but I will take that furnished by Mr. Stoddart, who relates in one of his works that he once captured a Pike with a gorge hook, and transferred him, with the tackle still in his throat, to the creel, from whence, however, he managed to jump out into the water, and almost immediately afterwards took a second bait and was finally basketted. In a subsequent letter to myself, Mr. Stoddart says:—'On one occasion I fairly "gagged" a Pike with my tackle, which he broke, and, although one of the hooks was in his upper jaw, and the other in the lower, in such a way that he could not open his mouth freely, he actually persisted in seizing the bait again and again, with such teeth as he could bring to bear.' Here, then, was a fish with a sharp barbed weapon fixed into certainly one of the most sensitive parts of its body, not only showing no symptoms of distress, but feeding with undiminished appetite. Surely we are justified in concluding from the analogy of the case—indeed, we cannot escape the conclusion—that such opposite phenomena argue an opposition in the accompanying pain equally radical?

Or take another instance, that of a sea fish, the Ling, mentioned by Mr. Crouch. 'I once,' he says, 'saw a Ling that had swallowed the usual large hook, shaft foremost, of which the point had fixed in the stomach, and, as the line drew it, it turned round, entered the opposite side of the stomach, and fastened the organ together in complicated folds; yet, having escaped by breaking the line, it survived to *swallow another hook, and be taken several days after.*' What does experience show would have been the feelings of a human being similarly circumstanced?

One more instance, which occurred under my own observation, and in the presence of several witnesses:—I was Perch fishing from a boat on Windermere, and in removing the hook from the jaws of a fish one eye adhered to it. I returned the maimed Perch (which was too small for the basket) to the lake, and being somewhat scant of bait, threw the line in again with the eye attached,—there being no other bait on the hook. The float disappeared almost immediately, and on landing the new comer it turned out to be the very fish I had the minute before thrown in, and which had thus *been actually caught by its own eye*! In a recent correspondence in the *Field* on this subject, several instances identical in character, and very nearly so in detail, with the above have been vouched for by the persons to whom they happened. On comparing the effects produced by these injuries to three of the most sensitive fish-organs, with the effects which would have followed in the human subject similarly wounded, I say the conclusion is irresistible that pain, in the sense in which we are conscious of it, is unknown to fish.

But it may be urged, perhaps, that if the laceration of the hook is not an actual cause of acute suffering in fish, the fight for life and its accompanying exhausting struggle are so. In reply, let me quote the testimony of Mr. J. A. Andrews, who was witness, in company with Captain Laurie, Mr. Knowles, and John Harris, fisherman, of Weybridge, to the following incident:

'When fishing in Ireland during the present year (1862) I was witness to an extraordinary occurrence—viz., a Salmon which had been hooked, and played for a considerable time, taking a *second* fly. Mr. Knowles was wading on one side of the Galway river, and Captain Laurie was fishing from the opposite bank; the former hooked a Salmon, and had played it for some minutes—at least forty yards of line being run out—when it suddenly made a dart across the river, and took Captain Laurie's fly. Supposing that he had hooked the fish foul, Captain Laurie gave line, and the fish was eventually gaffed on Mr. Knowles's side of the river, when it was found that both flies were hooked well in the inside of the mouth in the same corner.'

This account was kindly written at my request by Mr. Andrews, and by me published in the *Angler-Naturalist*, with the following additional anecdote, mentioned by Mr. Andrews in a subsequent note: 'I have since heard of an occurrence still more singular than that before alluded to—viz., a Salmon taking two shrimp baits, *the second when actually beaten, and just coming under the gaff.*' The foregoing are, I believe, the only two instances of the kind on record, but they are amply sufficient to prove my point— the other examples might easily be multiplied *ad infinitum*, but I have selected those given because they are authentic, and vouched for in each case by living authors able to verify their own statements if necessary.

Thus much as to the analogy—or rather absence of analogy—existing between human and fish organization. The fact is, however, that, even in the typical race itself, whose structures are, of course, in every case similar, wide gradations of capacity of suffering occur. Of the existence of such gradations, though they are not perhaps very generally recognised, recorded facts leave no room for doubt. Take, for example, the Indians of North America—by no means a very low type, but whose minds and bodies are both rude and uncultivated. We find that the American Indian will undergo, almost with apparent indifference, tortures, to our ideas, the most excruciating. But this, perhaps, it may be said, is mere stoicism, the result not of a want of education, but of a very high degree of education in a peculiar line. I do not think the argument is tenable, but take a case in which, were the Indian's organization equally susceptible to pain, no mental effort or education could prevent the overt physical symptom—I mean the amputation of a limb. It is stated, and I believe is generally admitted, that the American Indian will undergo the amputation of a limb, which in a European is usually attended with complete prostration, without any corresponding physical symptoms. Or take another instance from a still less educated race—the aborigines of New South Wales. A favourite amusement with these people is what they call 'waddying,'—that is, for two warriors to belabour each other upon the back of the head with heavy clubs, or 'waddies,' used with their whole strength, blow for blow, alternately, until one or other is stunned, or, what more frequently happens, is tired of the amusement! Does not this point to an insensibility to pain, suited, no doubt, to their lower development of nervous system, but incomprehensible to us! In short, without multiplying instances, it is evident that sensibility or insensibility to suffering, even in creatures of similar organization, is a matter depending, in a great degree, upon their relative cultivation and nervous sensibility, and that it is capable of every gradation of development, from the tenderly nurtured European lady, who will hardly put her foot to the ground 'for very delicateness,' to the shoeless denizen of Africa or Australia, whose indurated and horny cuticle treads upon the cactus and prickly pear without being conscious of inconvenience.

But the European is not so much raised above the savage as the savage is above the highest of the brute creation, and therefore it is reasonable to assume that even the most intelligent and highly organized of the one class suffers immeasurably less pain from a given injury than does the least cultivated of the other; and (a conclusion to which physiology also clearly points) that the descending scale is continued in a constantly decreasing ratio amongst animals themselves, until at last we come to creatures, like Lamarck's Hydra, which can be positively cut in pieces, not only without any extinction of life, but, on the contrary, to the end of creating a fresh existence out of each dismembered portion! The very distinctions between animal and vegetable have, in some of the lowest forms of organization, hardly yet been definitely settled by science, such 'thin partitions do their bounds divide,' and it does not require much consideration to perceive that there is probably every gradation of sensibility to pain, and to pleasure also, between the lowest and highest types of animal life, and that of such types the fish is one of the least capable of either.

With regard to the creatures used as bait by the angler, almost all the foregoing arguments, both physiological and analogical, hold good, and their capacity of suffering may be assumed to be of the lowest. A wasp will fly away leaving the greater part of his body behind him; butterflies or moths impaled and apparently writhing upon contiguous pins will frequently commence or continue the act of coition; worms, which possess a similar nervous system, I am told will do the same; and the organization both of these and of gentles or maggots, and of grubs generally, is of the most rudimentary description.

Fish, then, I repeat, as well as the baits commonly used in catching them, *have no capacity for feeling pain in the sense in which we are conscious of it*; that is my sincere conviction, arrived at after some thought and a good many opportunities of judging; and I hope the facts and arguments on which I have reached that conclusion will carry the same conviction to the minds of my brother fishermen, and thus enable them to enjoy their sport without compunction.

THE END.

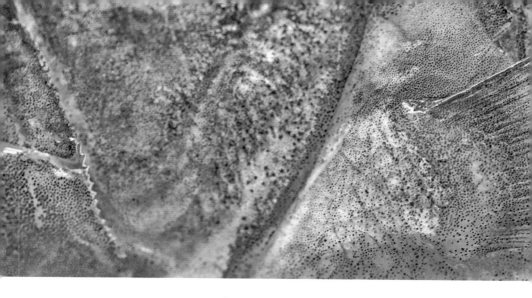

Chapter 1
Milestones Of
Fish Feel Pain
Science And Debate

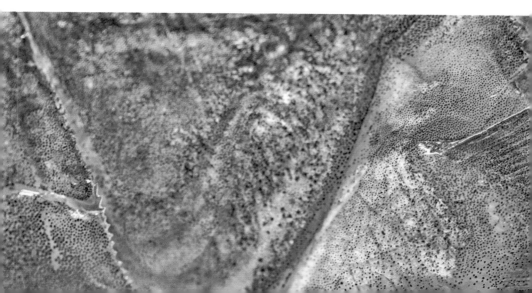

1870

Author: H. Cholmondeley-Pennell

Contribution: *Can Fish Feel Pain?*

Cholmondeley-Pennell had already anticipated most contemporary issues a century and a half ago:

1. The cruel angler: Angling can't be cruel provided the death or the presumed suffering of the fish originates in a good reason. 'Unnecessary pain or intentional waste' are to be avoided 'on fish or any other animals'.

2. In Cholmondeley's day FFP-advocates would point to 'nerves of sensation' and say that the 'organization' of fish enables them to experience pain. Cholomondeley shows that fish pain is a 'reflex action of the nerves'.

3. Cholmondeley also highlights the manifestly different fish brain and notes the absence of similarity between 'human and fish organization'.

1870–2001

Various minor studies and rhetoric here and there but nothing of major significance. However worth mentioning are the Royal Society for the Prevention of Cruelty to Animals (RSPCA) sponsored Medway report (1980) and a study by Verheijen and Buwalda (1988): *Do pain and fear make a hooked carp in play suffer?*

2002

Author: James D. Rose

Contribution: *The Neurobehavioral Nature of Fishes and the Question of Awareness and Pain*

The essence of Rose's findings is that fish do not have the brain for the pain experience. This translated into the popular slogan: *'No brain, no pain.'* Just as important as the no pain angle is the conclusion that 'awareness of fear' is 'impossible' for fish.

2003

Author: Alexander Schwab

Contribution: *Hook, Line and Thinker – Angling and Ethics*

Schwab attempted to take stock of the debate in terms of contemporary ethics and science up to that date. He concluded there was no pain and no cruelty, and if there was any doubt, then individual conscience (informed decision) should be the guide.

2003

Author: Lynne U. Sneddon, Victoria A. Braithwaite and Michael J. Gentle

Contribution: *Do fishes have nociceptors? Evidence for the evolution of a vertebrate sensory system*

This was the headline-grabbing game changer. The world of fishing (all kinds of fishing) hasn't been the same since. After studying the 'adverse behavioural and physiological effects' the researchers concluded that rainbow trout feel pain and mention that further research is required into the 'mental wellbeing of this species'.

2006

Author: Felicity Huntingford et al.

Contribution: *Current issues in fish welfare*

Huntingford et al. assume that fish have feelings and experience fearlike states. The authors – among them the professional utilitarian ethicist Peter Sandøe – defined fish welfare as the 'absence of suffering' and stated that:

> Wild fish naturally experience a variety of adverse conditions, from attack by predators or conspecifics to starvation or exposure to poor environmental conditions. This does not make it acceptable for humans to impose such conditions on fish...

If fish have feelings and they are hurt by fishing (or any other activity negatively impacting fish e.g. boating) all kinds of fishing are morally not acceptable. Once all fishing has been banned there are no more welfare problems caused by anglers.

2008

Author: Newby and Stevens

Contribution: *The effects of the acetic acid 'pain' test on feeding, swimming and respiratory responses of rainbow trout (Oncorhynchus mykiss)*

Newby and Stevens attempted to repeat the Sneddon et al. 2003 study but found several different results. Most notably, they could not replicate the famous rocking behaviour, which most likely was due to recovery from anaesthesia in Sneddon's fish. Sneddon et al. denied this and insisted that gravel on the tank bottom is a precondition for the rocking behaviour.

2009

Author: Robert Arlinghaus et al.

Contribution: *Contrasting pragmatic and suffering-centred approaches to fish welfare in recreational angling*

This contribution side-stepped the still very much in vogue philosophical viewpoint of Huntingford et al. and focused on

> objectively measurable variables of impaired fish welfare (e.g. physiological, behavioural or fitness indicators) and does not question recreational fishing on moral grounds.

Fish welfare is not based on feelings but on welfare science looking at the facts regarding the general state, fitness and health of fish.

2012

Author: James D. Rose et al.

Contribution: *Can fish really feel pain?*

Over the years since 2003 the number of studies purporting that fish feel pain had steadily proliferated. This contribution offered a detailed analysis of the scientific merits of the major FFP studies up to that moment in time. Its analysis identified serious flaws in the FFP studies especially in the area of replicability of experiments and the feasibility of some of the neurological interpretations.

In Sneddon's experiment, for example, venom, acetic acid and a saline solution were injected in the upper and lower lips of the trout experimented on. The 'saline fish' were 'stressed' due to handling and procedure but did not show the famous 'rocking motion' taken to be 'reminiscent' of primates in poor welfare nor did the 'saline fish' perform 'anomalous behaviours'.

Rose et al. observed pointedly:

> Embedding a fish hook is comparable with the mechanical tissue damage caused by embedding a needle of similar size, but without the saline injection…

Nevertheless the Sneddon group of researchers assert that their findings are of relevance to angling although their own data do not support such a claim.

Rose et al. concluded their analysis of the FFP science:

> Overall the behavioral and neurobiological evidence reviewed shows fish responses to nociceptive stimuli are limited and fishes are unlikely to experience pain.

2014

Author: Lynne U. Sneddon et al.

Contribution: *Defining and assessing animal pain*

In a direct response to Rose et al. in 2013 and later, 2014, in the article *Defining and assessing animal pain,* Sneddon et al. rejected any shortcomings. More significantly: they proposed a tick box system in order to establish whether or not an animal (fish) feels pain. This tick box system bypasses human pain definitions which, according to Sneddon, are 'not precise enough' and instead they focussed on behavioural changes which they counted as *'painlike behavior'* or *'consistent with the idea of pain'*. NB: 'idea'!

On that basis they determine whether or not an animal feels pain. The mainly behavioural key principles and criteria of this approach have been widely adopted by scientists and philosophers with a leaning to animal rights. Its critics claim that Sneddon et al. are shifting the goalposts on how they define pain, to suit their case.

2015

Author: Brian Key

Contribution: *Fish do not feel pain and its implications for understanding phenomenal consciousness*

This article published in the journal *Biology and Philosophy* is to date the most concise summary of the view that fish do *not* feel pain.

Key puts the emphasis on the lack of the 'neural circuitry' in fish which is required for the sensation of pain. To this day it draws considerable ire from those who believe in fish pain and animal rights. Key applied stringent logic and reached clear conclusions of uncomfortable clarity, to the consternation of the FFP camp.

2016

Author: Culum Brown

Contribution: *Fish pain: An inconvenient truth*

This contribution was a reaction to an invited target article by Brian Key in a new publication (2015) *Animal Sentience* where Key reiterated the thrust of his arguments.

Brown's article is representative of a different perspective that has significantly evolved since Lynne Sneddon's original research. The author states that FFP is a '*hot political topic*'. Brown notes that Key's target article:

> ...has so far elicited 34 commentaries from scientific experts from a broad range of disciplines; only three of these support his position. The broad **consensus** [bold added] from the scientific community is that fish most likely feel pain and it is time governments display courage enough to act.

Key's article was characterised by his opponents as 'simplistic', 'misleading', 'selective' and 'outdated'.

Brown is emphatic on the 'consensus' angle:

The vast majority of commentaries – experts in a wide spectrum of relevant scientific and ethical specialities – accept the accumulating, multidisciplinary evidence that it is likely that fish feel pain; moreover, even in the case of the fence-sitters (including optimistic agnostics), it is quite apparent that the precautionary principles apply, given the monumental number of fish killed each year in commercial fisheries.

It is worth noting that the 'vast majority of commentaries' refers only to readers of *Animal Sentience,* the journal where Key's invited target article was published. *Animal Sentience* is supported by The Humane Society of the United States which promotes animal rights. Its editor is a vegan animal rights activist.

Of course there have been other contributions of outstanding quality and impact including, for example, Victoria Braithwaite's *Do fish feel pain?* published in 2010. Yes, they do, she claims, in her accessible and well presented book. This publication has in the meantime become a classic of FFP-literature and is referred to and quoted on hundreds and hundreds of web pages and blogs affirming that fish feel pain.

Publish pain or pain will perish

In the last century, science has made some spectacular advances and progress seems to be accelerating at an ever-increasing pace. Just think of the astounding feats and real progress made in medicine, the internet and AI. The internet, AI and the advances in medicine are real.

Unreal on the other hand is the output in scientific literature. Estimates vary but Uncle Google tells us that there are currently around 30,000 journals publishing somewhere around 23 million articles a year. Many of them are probably only read by the author. One reason for this situation is the pressure on academics to publish in order to advance their career. Another reason is the possibilities of electronic media. It's big business for reputable publishers – and likewise for the fraudsters who run predatory or pseudo journals.

In either case it is an excellent business model: authors submit their articles and pay to be published, with the peer reviews providing the quality control. The peer review process is the gold standard of academic publishing. Reviewers traditionally work for free – it's part of the academic ethos – and it's time-consuming work if you're serious about it. When an article has passed the reviewing stage and the editor approves, it is published electronically. Result: low cost, high margins and ideally high quality. Ideally.

The reality is apparently that there are (as of 2022) an estimated 8,000 pseudo journals out there which publish 2.5 million papers per year. These pseudo journals skip or fake the peer review – all they are interested in is the fee the author pays which can range from USD 150–6,000. The result: low quality, high output, phenomenal amounts of money changing hands.

FFP is a scientific question as well as a 'hot political topic'. As such it concerns not just science qua science but philosophy, sociology, psychology, history, law and more. Lowering the standards or shifting goal posts facilitates desired outcomes. If you can flood the marketplace with papers that back up your agenda, this can change the perception of the issue, if only by claiming that a majority of scientific papers back the FFP claim. That applies to all controversial causes. Research these days is done electronically and with more and more help from AI and the sheer number of published papers, even though nobody has read them, will become useful: X number of studies say Z therefore Z. It's a way of composting trash. Any plain nonsense can get published under the flag of 'science'. Worse than that, it is open to fraud: somebody submitted a fake article written by AI under the name of a recently deceased FFP author and it got accepted and published.

This large-scale publishing scam undermines trust in science. These fake articles wear down the authority of science just as the internet and social media have diluted and devalued information. There is nothing new under the sun. Think of salt. Salt once was a very valuable commodity. Roman soldiers were paid in salt, it was a currency and the Latin word *sal*, meaning salt, is at the root of the word 'salary': to this day you can be 'worth your salt'. However, salt now is everywhere and cheap as chips. It has been devalued. And it's said to be bad for all sorts of things like blood pressure!

Mind you, that can change again — which brings us back to science. Judge by your own experience. One week you read in the news that coffee drinkers are the least likely to get cancer and the following week another study/paper tells you the exact opposite. It seems that for practically every study which says X there is another one saying Y, both claiming to be 'science'. For the non-scientist this is more than confusing, especially if it's a matter of real concern that you must decide on. On what basis will you make your choice between X or Y?

Of course, you would like to understand the matter but what if you need to make sense of a study dealing with, for example, the role of tricarbolic acid in mitochondrial signalling? You could always ask Uncle Google to help you but, if you're not a specialist in that field, it's nearly impossible to understand the content your search shows up.

You and I are at the mercy of those who produce and feed us the information. It looks as if we are caught in the middle of something like a simultaneous explosion and implosion of the production and distribution of information. Perhaps the situation is still so alive and dynamic that we haven't found a way yet to deal with it. That is not a very satisfactory conclusion, but at the moment there is no silver bullet here.

I can't resist finishing this section on an amusing note. In 2018, the journal *Animals* published an article with the promising title: *Can Donkey Behavior and Cognition Be Used to Trace Back, Explain, or*

Forecast Moon Cycle and Weather Events? It's all about 78 Andalusian donkeys who were volunteered but there are also fish involved:

> Slight barometric pressure fluctuations have traditionally been reported to promote behavioral and feeding activity in fish. Fishers usually relate slight changes towards high pressure to clear sky occurrence during which fishing is medium to slow as fish may slowly be in deeper water or near cover. These trends progressively invert when there is falling pressure, the best attributed timing for fishing during degrading weather when fish are more active, may support our results.

This is reminiscent of: 'The two best times to fish are when it's raining and when it ain't.' However, back to the donkeys:

> Conclusively, donkeys can be used as an environment informative sensitive tool and may therefore predict and register slight human-unappreciable climatic variations to which they may behaviorally adapt beforehand.

Impressive as this conclusion is, there is some truly stunning material in the ethics statement for the article: 'All subjects gave their informed consent for inclusion before they participated in the study.' If it's a hoax, it's priceless and if it's not – you can't weigh it up in gold.

Believers and sceptics

Pain, fish pain, is absolutely pivotal to animal rights. Without it the entire philosophical and ideological house of cards collapses. I discuss this later on in detail – for the moment just take it as a given. Fish pain research and debate before Lynne Sneddon seems somehow bland. Nobody got really excited about it whereas Lynne Sneddon's study went viral from the start. Why would that be? Some of the earlier non-Sneddon studies came down heavily on the side of fish pain and identified anglers as 'barbaric', but the world's media weren't interested.

There are, I think, two reasons for Sneddon's media success: first the novelty involved in Sneddon's study (*nociceptors – see page 42*) and secondly – and perhaps more importantly – the social climate. What do I mean by that?

The modern animal rights movement as we know it today starts with Peter Singer's book *Animal Liberation*. Seemingly out of nowhere *Animal Liberation* (1975) created the contemporary animal rights movement practically overnight. Like its succeeding work *Practical Ethics* (1979) it became the staple diet of college and university curricula all over the First World (this is still the case today). Literally millions of students were exposed to Singer's ideas and it heightened awareness of issues relating to animals. Hugely popular stories like *The Rainbow Fish* (1992) and *Finding Nemo* (2003) played their part too by elevating fish to quasi human status. For a variety of reasons the First World was ready for the FFP-message. It is however important to note that animal rights is part of an ethical package tour (Singer: 'animal liberation is human liberation') involving racism, gender and all those related issues. In an ideal world 'Science' like 'Justice' would be blind (i.e. impartial). This is not case though: neither science nor justice is blind. Scientists, like the rest of us, have their history, their experiences, their preferences and opinions. The fact that they hold the views they do isn't necessarily tied to their field of science. You can engage against nuclear proliferation without being a nuclear physicist or a scientist at all. It's different, of course, if as a scientist you streamline your findings to fit your mission. One of the leading pro-FFP scientists who believes that fish feel pain and that they suffer had on her website the famous but fake Gandhi quote: 'The greatness of a nation and its moral progress can be judged by the way its animals are treated.' There couldn't be a more phoney and misanthropic statement than that. The point here however is that if you operate under a pre-held, guiding principle, it might just taint your findings a tad.

In the twenty years since Sneddon's article, the activism component in the FFP issue and the momentum and impact of the animal rights movement has increased exponentially. Wikipedia was established in 2001 and certainly from 2014 onwards the fish pain page has been substantially restructured by activists who erased much of the previously objective and balanced discussion that used to be there. Today it is protected by highly motivated guardians who realised long ago that the Wikipedia page is the first thing that comes up when an inquiring mind types the words 'pain in fish' into Uncle Google. Not only that: while in 2003 it was just the anglers on the hook, the scope now has significantly widened and includes *all* forms of fish use.

Just as there are 'FFP believers', there are also 'FFP sceptics'. The sceptical scientists take issue with some of the findings of the believers in the areas of the experiments themselves, the methodology and the interpretation of some of the results. It's all intricate and complex and not easily accessible for the non-specialist scientist. We have to take either the believer's or the sceptic's word for it. Instead of guidance by science, the interested angler and consumer of fish products gets conflicting information. The situation is not helped by a kind of scientific trench warfare that has been going on between believers and sceptics since 2003, with no end in sight. However, judging by media echo, the believers have the upper hand at the moment.

The point to take on board is that the FFP issue has, over the years, morphed into a high level cluster of scientific, political, ideological, philosophical, economic and ethical debate, the different strands of which can be separated only with great difficulty, if at all. In fishing parlance, it's a hopeless tangle. This is the result of the educated and clever agitprop of the animal rights movement, supported intentionally or unintentionally by FFP-believers.

The 'consensus': Fish Feel Pain (FFP)

Since the inception of FFP in 2003 twenty years of additional fish pain research has been inflicted on fish. However, there is a glimmer of hope that their agony in the science labs will soon be over. As one commentator claims: 'Scientists have reached a consensus on the issue of fish pain, concluding that their suffering is real.' Consensus? What exactly is meant by 'consensus'?

A nice specimen of 'consensus' is *The Cambridge Declaration on Consciousness*. What's that all about? In 2012 a group of eminent neuroscientists published *The Cambridge Declaration on Consciousness* in which they stated that non-human animals possess the 'neurological substrates' for consciousness.

Whatever the merit and whatever was intended by this declaration, for philosophers and animal rights activists this is manna from heaven. It can easily be construed, for example, as:

There is an emerging consensus that current evidence supports attributing some form of consciousness to other mammals, birds, and at least some cephalopod molluscs (octopuses, squid, cuttlefish).

The Cambridge Declaration on Consciousness is yet another 'consensus' frequently invoked by FFP/animal rights activists. I wonder what the not-so-eminent neuroscientists around the world make of this declaration? It is what it is: a declaration and as such, a political statement, rather like a 'declaration of independence' or a 'declaration of war'.

'Consensus' is 'an agreement among a group of people'. There surely is a group of people – the self-appointed leaders of that field – among which the consensus is that the earth is pretzel-shaped (quite an attractive idea, really). Not so long ago there was a very vociferous scientific consensus ('group of people') claiming that we are heading for a new Ice Age.

In the past the consensus was, for millennia, that the pretzel was at the centre of the cosmos (geocentrism). Consensus regarding FFP is what many philosophers and the animal rights movement rely on. But as outlined in the milestones at the start of this chapter, there is certainly no consensus in the scientific community regarding the FFP question.

A fish called Rudy Rainbow

Nociceptors first. What is a nociceptor? A nociceptor is a nerve cell which picks up noxious stimuli. Nociceptors are the first sensory link in the pain chain: If you burn your finger on a hot plate the nociceptors in your finger send 'data' to your spinal cord which triggers the reflex withdrawal. After the reflex a message is sent to the brain, which in turn processes the information to pain and you go 'Ouch'! Note that nociceptors are not 'pain detectors'. The messenger is not the message. Pain is created in the brain. Not in any brain but certainly the human brain. Human pain is the frame of reference for pain in fish because there is no other.

The following might seem pedantic but that 2003 header of Lynne Sneddon's article, '*Do fishes have nociceptors? Evidence for the evolution of a vertebrate sensory system*', merits a closer look.

Consider the following: one trout is a fish; two trout are two fish. A snapper is a fish; a snapper and a gurnard are two fishes. Is the plural of fish: fish or fishes? The word of choice in the article's heading was 'fishes' meaning presumably all 30,000 species of fish in the world. That is the obvious interpretation which the world's media and practically everybody else took on board. The message had gone out that ALL fish feel pain in the FFP debate.

This has required some backtracking. In a letter to *The Guardian* in 2021 Lynne Sneddon states:

> I was the first to identify the existence of nociceptors in a fish, the rainbow trout, in 2002 [published 2003]. These are specialised receptors for detecting injury-causing stimuli, and their physiology is strikingly similar to those found in mammals, including humans. Since then, my laboratory and others across the world have shown that the physiology, neurobiology, molecular biology and brain activity that many fish species show in response to painful stimuli is comparable to mammals.

That is a very interesting message and the devil is in the detail. Sneddon is now referring not to fish of all species which feel pain, but 'many fish species', and out of the 30,000 odd 'many' she singles out one species which just happens to be very popular with anglers.

The FFP claim is species specific and relates to individual fish e.g. Rudy Rainbow. Note that the nociceptors (*triangles in photo below*)

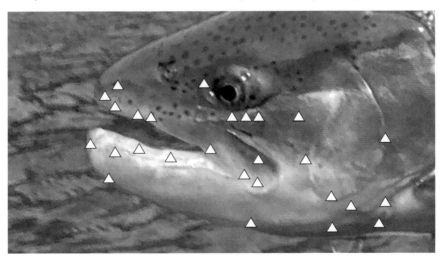

Above: Nociceptor cells are the first contact point of the process that creates pain in the brain. In the case of the rainbow trout they are located mainly around the mouth area. They detect injury (noxious stimulus) but not pain! Some fish do and others don't have nociceptor cells.

are conveniently placed around the areas where a hook would penetrate. It is perfectly reasonable – if wrong – on the basis of the available evidence to claim or assume that rainbow trout, Rudy Rainbow and friends, feel pain. But the claim that all 30,000 species of fish feel pain is unreasonable.

Whatever animal pain we talk about, it always refers directly or indirectly to the human experience. In 2003 Lynn Sneddon seemed to think so too:

> If a noxious event has sufficiently adverse effects on behavior and physiology in an animal and this experience is painful in humans, then it is likely to be painful in the animal.

However over time the definition of human pain in relation to fish has seemed more and more inadequate.

In 2014 Lynne Sneddon et al. suggested a novel box-ticking approach which would bring precision into determining whether or not an animal (fish) feels pain. At the heart of the box-ticking system are mainly behavioural criteria (reactions to presumed painful stimuli) like protective behaviour, altered behavioural choices/ preferences, rubbing, paying a cost to avoid a noxious stimulus or reduced responses after use of analgesic drugs. Only if all boxes can be ticked, the animal (fish) feels pain. Among the non-behavioural items to be ticked we find 'Receptors for analgesic drugs'. This surely is just an unfortunate choice of words because it means over millions of years, fish evolved in anticipation of pharmaceutical discoveries i.e. analgesic drugs. As you would imagine not everybody buys into the box-ticking approach.

One main criticism is that on closer inspection none of the criteria either by themselves or together require pain. The pain bar is set so ambiguously and so low that with a little bit of tail wind insects can easily fly over it. This is the reason why the box-ticking approach has been characterised as 'shifting the goal posts' and lowering the evidential crossbar. Besides, however hard you try you can't escape the human angle. All the criteria brought into play make sense only in relation to the human pain experience. A different, not scientific, criticism is that any pain human or animal is subjective which is why there is a certain fuzziness to it. You can't be more precise than the subject matter actually allows.

However, we all have a trigeminal nerve and so have rainbow trout. The trigeminal nerve manages the information around your face and mouth. Of all the cells involved in this complex business, 80% of the mammal trigeminal nerve cells are so-called C fibre nociceptor cells. Their job is to message you, for example, that you have an excruciating toothache. In contrast to humans, the management of the trigeminal nerve in rainbow trout, according to Lynne Sneddon's experiment, employs only 4% of C fibre nociceptor messengers. These C fibre nociceptor cells have also been found in other bony fish but so far none have been found in elasmobranch fish like the shark or the manta.

There is another type of nociceptor cell which is important: the A delta fibre type. If you feel a sharp pain like when you catch your finger in the door, the A delta nociceptor cells are the messengers transmitting the quick news. Later on, if your finger was hit quite badly, the slower C delta fibre cells will confirm the bad news when you feel a dull throbbing pain. Many fish have these A delta fibres and they trigger the immediate escape and avoidance responses. However, according to Lynne Sneddon these A delta fibres are a kind of super fibres because her research shows '... that trout A delta fibres act in the same way as mammalian C fibres'. That is a courageous speculation to say the least, and so far it has not been independently confirmed.

My body is my brain?

Probably the most quoted and commented part of Lynne Sneddon's experiments on rainbow trout is the 'rocking' behaviour displayed by trout after an injection of bee venom. Rudy Rainbow had a massive dose of bee venom injected into his lips, after which he stopped feeding

© Merlin Unwin

and showed the now famous rocking behaviour. This is taken by many commentators, scientists and philosophers as straightforward evidence that fish feel pain.

But is it? Could the bee venom have simply affected Rudy Rainbow's sense of balance and made him wobbly? For example, for many people if they enter a room with an overpowering smell, they don't feel pain but they become dizzy and unsteady on their feet. However, with a bit of fresh air they start to recover.

Let's also consider that 'massive dose', because if a proportionate amount (in proportion to body and brain size) of bee venom was injected into your lips you would stop feeding forever. Rudy Rainbow took only three hours to resume feeding. How's that for resilience?

Whatever is made of the rocking story, a major bone of contention is what information fish are able to obtain (remember the 4% C fibres) and what they are capable of doing with it. The fish brain features no neocortex. In humans the neocortex is broadly the pain centre hence the popular formula 'no brain, no pain'. That much is true, but the brain is only half the story. Just as important is the entire 'neural circuitry' of the organism 'fish'. The believers say with some justification: focus exclusively on the brain is not good enough. They on the other hand – or at least some of them – completely play down the surely important differences between the human and the fish brain, which have both evolved for different tasks.

As Matthew Chalmers puts it:

Seeing as fish exhibit all the same physiological and neurological reactions to noxious stimuli that other sentient animals do, one would need to have a particularly strong reason to think that their brains are too simple to experience pain. The absence of the neocortex is insufficient to establish this, and science is progressively uncovering the way that consciousness and subjective experience are not solely a feature of mammal-type neurology.

He concludes: 'Fish feel pain through different systems'. In other words their body is their brain. The entire fish is a brain. The fish is not a fish but a brain. At least that is what I take it to mean. Again it is important to point out that what is said in this respect about Rudy Rainbow may not apply to Michael Manta-Ray either as an individual or a species.

In the wake of Lynne Sneddon's experiments and findings, heaps of experiments have been designed to support the FFP claim. These experiments tend to focus on avoidance behaviour and the learning ability of fish. Electroshocks and alcohol are the scientists' weapons of choice to show that fish – especially zebrafish – avoid areas in their tank where they were subjected to particularly unfriendly treatment.

There is some perplexity involved in the process: in order to treat the fish 'humanely' Sneddon and other FFP researchers anaesthetise the fish before injecting either venom or acid in the fish's lip. When the anaesthetic wears off the venom unleashes its full pain power. In order to show that fish feel pain, morphine is brought into play either by injection or added to the water and the 'pain-like behaviour' is either reduced or stops. This then is taken as an indicator that fish feel pain (for philosophers it is proof). It could, however, also be the case that the morphine just ends nociception. I just wonder whether it wouldn't actually be more 'humane' to drop the anaesthetic because it would be one severe attack less on the well-being of the fish.

Cats, dogs and humans would drop dead on the spot if proportionally the same treatment was meted out to them. I am actually amazed by the cold bloodedness of these researchers. They claim to know that fish feel pain in a similar way to mammals, yet aren't they practising vivisection for the sake of it, just to confirm what they already know? Of course, the answer given to my question would be: this is different because it's in the name of science, and you're not a scientist. Sure, but if fish do feel pain, whatever the purported aims of the experiments are, fish are nevertheless thrown into the torments of these tests by scientists. There is no end to the research agony in sight for these lab fish not least because there are 30,000 or more fish species to be researched.

Other types of experiments show the amazing abilities fish have in learning and navigation. A stunningly clever experiment was designed by scientists from the Ben Gurion University of the Negev. BBC *Science Focus* reported in January 2022: 'As it turns out, goldfish can take to driving a car like a duck to water.' Scientists designed a fish tank on wheels and the goldfish learned how to drive the aquarium around. Check out the video – great stuff!

Great stuff? Hold it there – didn't I just get a cheap laugh at the expense of a goldfish being coerced into motoring his tank in the right direction? The goldfish certainly didn't volunteer for this experiment. Driving a fish tank around is certainly not an expression of a goldfish's normal behaviour. In fact what we are witnessing is a sentient being forced to perform acts it would never dream of in its natural environment. According to Swiss Law this experiment would probably be a criminal offence because it violates the dignity of Gwendolyn Goldfish to an unacceptable degree. Nevertheless: well done, Gwendy! And lucky she was – she didn't land in the hands of a professional torturer (fish feel pain scientist) deploying all his skill to extract some information out of the poor fish regarding its pain and suffering.

For FFP activists learning and avoidance are irrefutable indications that fish feel pain and suffer: if we have the brain to drive a tank and the brain for pain and suffering and they have the brain to drive a tank, then they must also feel pain and suffering. We avoid electro-shocks and overdoses of alcohol because they make us suffer. Likewise fish avoid them because they are like little humans and it makes them suffer. That kind of popular equation is humanising fish to a ridiculous degree, unless of course you believe that the goldfish really is in fact human in a way – a cleverly disguised human?

Incidentally: artificial intelligence ChatGPT has remarkable associative learning abilities but that doesn't make the computer feel pain. Directly linking learning and avoidance behaviour to the ability to feel pain in fish is saying that the fish brain is a miniature human brain, which it isn't. And if it were, the fish experience would be a scaled-down human experience. Counter-factual as it is, for the ardent FFP believers, activists and philosophers this doesn't matter.

Has there been really something substantially new – irrefutably, factually new – since 2003 about the fish brain and its workings since 2003? The answer seems to be 'no', and believers and sceptics are as divided as ever. Wave upon wave of science and debate followed Lynne Sneddon's findings and observations. The debate is scientifically and philosophically intricate, convoluted and for a non-professional almost impossible to follow in all its ramifications. However, at some stage all the involved scientific and lofty philosophical discourse has to be couched in terms you and I can understand because the practical consequence (ban of fish use, restrictions) of the debate is of public interest.

The livelihood of 3 billion people worldwide depends on fish and fishing!

The most famous and influential philosopher and father of the animal rights movement, Peter Singer, has a ready-made solution for everybody:

'What is the ideal next step in regard to how we use fish?'
His answer: *'The ideal next step would be to stop catching them and eating them.'*

This is so deeply misanthropic and ugly it beggars belief. Nevertheless it is consistent with applied animal rights (*see page 102*). I want to emphasise we are not talking of any esoteric fringe figure but someone considered a First World leading intellectual! This is mainstream and headline stuff for the likes of the *New York Times* or in Britain, *The Guardian*. It is mainstream to believe that: 'There is no reason to say humans have more worth or status than animals.' A fish because it can feel pain and suffer is like a human or the human is like a fish.

Peter Singer backs up his position on fish, of course with reference to the FFP believers' science:

> On animal sentience, we now have strong evidence that fish too can feel pain. There are also good reasons for thinking the same of some invertebrates – the octopus but also lobsters and crabs. How far sentience extends into other invertebrates is unclear.

Think twice before you dismiss this FFP issue and animal rights as something which won't affect you. More on Peter Singer later on.

We haven't got the brain for it!

Am I barking up the wrong tree? Why focus on pain in fish? If as it is often claimed they are so intelligent, capable of learning and performing all sorts of tricks, why should they not be capable of feeling pleasure too? That is exactly what an animal rights website suggests:

> Although anglers and sport fishermen claim that fish lack feelings in order to justify impaling fish on hooks, violently pulling them from water and mercilessly suffocating them, a series of behavior tests have shown that instead of a merely reflexive response to pain, fish possess a conscious awareness of pain. Experts believe that fish are not only capable of feeling pain, but also pleasure – oxytocin, a 'feelgood' hormone, has been discovered in fish.

This conjecture has been radically debunked by recent research. This is no deterrent to FFP philosophers, however, because no research is important enough to persuade them to think differently nor to cease moralising on the basis of ignorance. One philosopher has it – seriously – that not only must we not cause pain in the world of animals but we must also create situations to bring pleasure to animals. So a new set of challenging questions rise on the horizon: what is fish pleasure? Is it the same for all fish e.g. barracuda and mackerel? Where in the fish brain does pleasure happen? Do fish smile (often consistent with pleasure)? And so on. Be that as it may it doesn't answer the basic question: what is pain?

The IASP (International Association for the Study of Pain) defines pain as 'an unpleasant sensory and emotional experience associated with, or resembling that associated with, actual or potential tissue damage.' It then expands this with six additional points two of which bear on the FFP question.

- Pain is always a personal experience that is influenced to varying degrees by biological, psychological, and social factors.

- Pain and nociception are different phenomena. Pain cannot be inferred solely from activity in sensory neurons.

Instead of arguing over whether fish feel pain,
why not engineer fish that definitely feel pain
and be done with it?

This is about human pain and human pain only because that is the only meaningful way we can talk about pain. We humans are the only beings that can talk about human pain. I emphasise 'talk' because my toothache is an absolutely unique experience to me. Nobody can feel my pain but me. However, since humans have similar (not identical!) pain experiences they can communicate about it, sympathise and find ways to fight it. I can't help remarking that it would be nice to have another 'International Association': the International Association for the Study of Pleasure.

One way humans communicate their pain experience is with the help of the pain scale whereby you assess your experience on a scale from 1 to 10. As far as I am aware, no one has yet explored the intensity of fish pain. What a challenge that would be! Just consider how two human beings can have completely different intensity experiences for the same painful event. Or to look at it a different way, the same amount of a painkiller doesn't do a thing for one person yet works wonders for another. Age, health, psychology and the weather all play their part in that uniquely individual experience. Pain is not just 'pain' but a complex process in an organism. Note that in the case of humans, pain is almost always specific: we invariably point to an area, source or causes (except, of course, in the case of *Weltschmerz i.e. world-pain*). Likewise we characterise pain, for example, as sharp, dull or throbbing in order to communicate what is happening in the organism that is you.

Let's go back now to Rudy Rainbow in Lynne Sneddon's experiment who is said to feel pain. No more information – just 'pain'. It is a legitimate question to ask whether or not this is any information at all. The pain would be fishy pain of course. But what does that mean in terms of intensity, and how is it characterised? Is it the same for male and female fish? For Rudy Rainbow it might be on the fishy pain scale at level 1, where you wouldn't talk of pain but more of a bit of an itch, a tingle or some other mild sensation. Or does the bee venom induce a level 2 hangover with a throbbing headache? Could this explain the rocking? The challenge for the FFP science is to establish what 'pain' actually means to Rudy Rainbow.

Surely if there is fish pain, there is also a fish pain intensity scale and there are also different pain characterisations. Assume now that fish feel pain and think of it in a practical way:

1. Rudy is fished out his tank with a landing net to be injected with bee venom. The clumsy researcher bangs Rudy's head on the tank edge. That could be a blunt 'ouch'.

2. The bee venom is injected (after anaesthetisation) and Rudy is put back into the tank and then recovers from the anaesthetic and gets the full blast of the venom and, somehow stimulated by the gravel in the tank, rocks along. Only in a tank with gravel could the rocking motion be observed. Surely that is not the same kind of pain as the tank edge 'ouch'.

All the events in this process are more or less hurtful (assuming that fish feel pain). To say that they are all the same pain doesn't seem to make much sense. If then you say that banging Rudy's head on the side of the tank doesn't register on Rudy's pain threshold and that pain is felt only with the bee venom injection, you are presupposing a pain scale of some kind – though you haven't got a clue what it is. The general statement 'fish feel pain' is meaningless to us because we do not know what it means in terms of the individual fish.

The quintessential difficulty with fish pain is the 'personal experience', the self, the consciousness, the 'I'. In order for there to be pain there has to be a kind of 'I' – a somebody of a kind with an idea of themselves. The mirror test is used to see whether an animal recognises itself in the mirror. It's of dubious value to the pain debate because some mammals (primates) pass it while others don't or can't be bothered with their own image. There was a lot of hullabaloo about a tiny cleaner fish 'passing' the mirror test. Cleaner fish are unique and extremely specialised coral reef fishes which spend their entire lives picking tiny parasites off other fish. They are innately disposed to pay close attention to any marks on fish bodies. Their self-recognition might thus be nothing else but them inspecting their bodies for tiny marks. Taking this 'pass' at face value would mean that the cleaner wrasse is on a higher cognitive level than a five year old child. There are apparently some significant cross-cultural variations regarding human children and the mirror test. First world, Western children pass the test between 18 and 24 months of age while others don't up to 72 months.

The famous primatologist Frans de Waal commented on the cleaner wrasse self-recognition:

> The mirror mark test has encouraged a binary view of self-awareness according to which a few species possess this capacity whereas others do not. Given how evolution works, however, we need a more gradualist model of the various ways in which animals construe a self and respond to mirrors. The recent study on cleaner wrasses *(Labroides dimidiatus)* by Kohda and colleagues highlights this need by presenting results that, due to ambiguous behavior and the use of physically irritating marks, fall short of mirror self-recognition. The study suggests an intermediate level of mirror understanding, closer to that of monkeys than hominids.

Although this mirror test angle is highly controversial I mention it here because it is often brought into play by animal rights activists to prove that fish are 'like us'.

Back to FFP. Assuming that the FFP claim applies to herrings too, then each of the herrings in a shoal has a sense of identity of himself. Harry Herring, for example, one in a shoal of hundreds of millions of herrings knows he is the one and only Harry Herring and that he is not going to be spared. The whale will gulp him too. In the darkness of the whale's digestive system Harry dies a painful death. However you twist and turn the pain story, there is no way past a 'somebody' because FFP is tagged to human pain and that requires a conscious individual. What kind of self is the fish? If somebody would say: it's 10% human – that would mean as far as self-awareness and mind are concerned, the fish are basically scaled down humans who happen to live in a watery environment, have a different body and scales to sustain their scaled-back 'human' brain.

The famous FFP scientist Victoria Braithwaite comments:

> Fish do feel pain. It's likely different from what humans feel, but it is still a kind of pain.

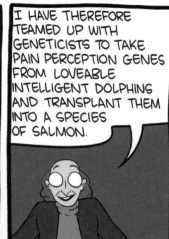
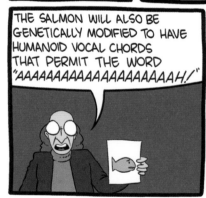

I can't help wondering whether that is a meaningful statement at all. I think the following is a fair representation of what is being claimed here:

- Humans feel pain
- Fish feel pain
- Fish pain is different from human pain
- Fish pain is still a kind of pain

I assume 'different' means different in kind and not different in degree because the latter would mean, again, fish are humans in disguise and vice versa. I don't think people who are seriously into FFP research consider that as an option. You can turn the statement upside down and take fish pain as a starting point which would be legitimate because it makes supposed sense to talk of pain in fish and fishes. If it is pain qua pain it should make sense in both ways. Otherwise it's non- or not-much-sense.

- Fish feel pain
- Humans feel pain
- Human pain is different from fish pain
- Human pain is still a kind of pain

Either way we don't know what 'pain' is – in fish that is. The nearest I have come to understand this reasoning is by playing around with the FFP-reasoning:

- Birds can fly
- Humans can fly
- Human flight is different from bird flight
- Human flight is still a kind of flight

That seems to make vague sense. However our flight is powered by jet engine whereas bird flight and all other 'flight' (bats, insects, seeds) relies on a completely different 'technology'. If you struggle with this analogy it's not because of the analogy but the difficulty of making sense of talking about something which we don't have clue about, that is, the experience of 'fish pain'.

There is another shortcoming of the flight analogy: pain (FFP) is different because pain is not something fish do. Rather it is something they experience and have the mental capacity for, and that is where the brain comes in again. There is no way around it. 'Pain is always a personal experience' the International Association for the Study of Pain says. We humans require not only the neocortex to feel pain but 86 billion neurons to get 'personal'. That seems awfully inefficient compared to the estimated 80,000 neurons of the zebrafish. That is if the FFP science and the reasoning supporting it got it right. Those fish brains would then be a kind of superbrain and indeed – why not? Quality tops quantity. No human can match the navigational achievements of a salmon, why should the powerful salmon brain not be capable of feeling pain which in comparison to navigation seems a banal achievement?

In a more stroppy spirit, under my understanding of the FFP reasoning I could assert that horses have wheels (fish feel pain):

- Cars have wheels (humans feel pain)
- Horses have wheels (fish feel pain)
- Horse wheels are different from car wheels (fish pain is different from human pain)
- Horse wheels are still a kind of car wheels (different pain it may be but still a kind of pain)

It still makes a kind of sense. After all horses undoubtedly move and although it's different from car movement it is still a kind of movement… However, I leave this angle here because otherwise I risk winning the Nobel prize for literature. I'll sum up by saying we can't possibly know what fish pain is if indeed there is fish pain. The reason: we haven't got the fish brain for it. At least I haven't.

However, there are legions of philosophers out there who are trying to work out such questions as 'What's it like to be a bat?' or 'What's it like to be a fish'? Just imagine the gargantuan dimension of the task: considering that you're not just exploring pain but the entire fishy experience itself. It can't be emphasised enough that pain is absolutely individual, and the same goes for the experience of yourself as yourself. You're floating in that unique bubble that is you. Your self is yours, and only yours and absolutely nobody else's. You can imagine, sympathise, empathise and do all sorts of human things, but in the end you can't really know what it is to be someone else, not even someone as close as your spouse.

Consciousness is a fascinating and absorbing field of study. As the *Stanford Encyclopedia of Philosophy* aptly puts it: 'Perhaps no aspect of mind is more familiar or more puzzling than consciousness and our conscious experience of self and world.' Not surprisingly there are numerous theories orbiting around the puzzle of consciousness. All of them come in different shapes and guises and meanwhile different schools of thought muse over gaps, inconsistencies, assumptions and new angles. In other words: it's work in progress and no end in sight. However sophisticated and intricate these theories might be, none of them manages to jump the interpersonal – let alone the species – barrier. The statement that fish feel a kind of pain which is not human pain is incomprehensible.

These highly contentious theories of consciousness are used by well-meaning activists to make believe that fish feel pain. Which fish: Zane Zebra? Gwendolyn Goldfish? Worse: on a fairly broad basis of ignorance and nonsense some philosophers and scientists feel entitled to demand a ban on fishing and fish use of all kinds. This is again where 'consensus' kicks in: knowledge and theory gaps are simply closed by manufacturing 'consensus'. However, the conundrum of the fish self, its consciousness, the intensity and character of the pain experience of Rudy Rainbow, Harry Herring, Zane Zebra and Gwendolyn Goldfish remains.

Beam me up, Scotty

Many things once seemed impossible or unthinkable but have since come true: landing on the moon, or the Berlin Wall coming down

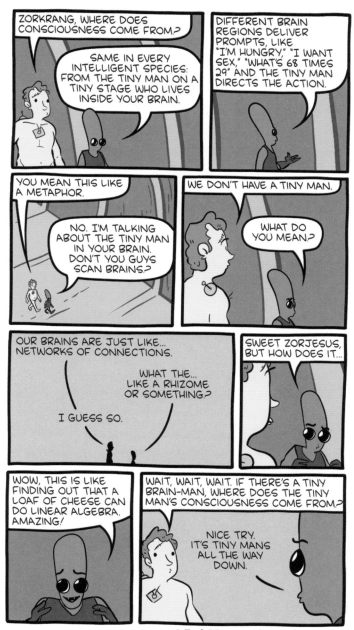

© Zach Weinersmith, smbc-comics.com

to mention but two. AI and super computers might open the door not only to mind-reading but might also give science access to the fish brain. The first step will be human beings linked over brain hoods and reading each other's minds. This will be made possible by AI analysing the physical and chemical processes of neurons firing around in the brain and then deciphering and verbalising these. This is assuming that straightforward thought is physical. There might be a problem with long term states like happiness or contentment or with a zone such as 'conscience'. Are they the same kind of physical occurrence in the brain as 'thought'? Whatever, it's probably all a bit like the weather: there is always something happening; it's a process producing different states.

Once science has tackled inter-human mind reading and the exchange of mental states the exploration of the fish brain will be easy peasy. Or will it? While it already seems no problem to visualise the neurons firing in the fish brain, it would still be a challenge to tap and translate the process into a meaningful human form. The problem would be the different neurology, the comparative lack of C fibres, the brain – the entire fish being the brain – which probably produces different processes which would leave FFP science in the same spot as it is now: 'Fish pain is different from human pain but still a kind of pain.' As shown above, this is a meaningless proposition.

But FFP believers of all persuasions need not despair: the impossible almost always happens. In an ideal world the scientist could be the fish and the human at the same time – and vice versa – which would mean the species barrier would fall and except for appearances there would be no species any longer. It would be the end of being human as we know it thus far.

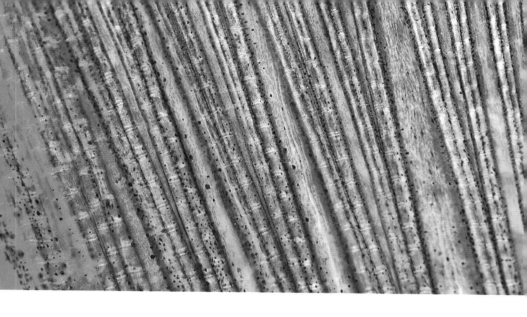

Chapter 2
Some Philosophical Definitions

The story in this chapter seems to drift even further and further away from fish and fishing. However, the further out we go the closer we get to the heart of the matter. It might seem trite but it is important to understand that recreational fishing takes place in a societal context. It is embedded in society and whoever calls the shots there commands the dos and don'ts. All fishing politics is 'hot' with real interests and that doesn't necessarily mean the interests of anglers. Quite the contrary in fact.

If you want to understand anti-angling, anti-hunting or any other 'anti' in regard to animal use by human beings there is no way around Peter Singer. His books sell in millions and are translated into all major languages. There is no college or Philosophy Department in the Western world which doesn't run a course on Peter Singer and this has been the case for decades. It is no surprise that his ideas trickled down from colleges and universities into everyday life. After all, teachers, journalists, lawyers, psychologists and sociologists, to list a few, are educated in colleges and universities and their curricula include a fair dose of Peter Singer's ideas.

In practical terms this means for example that the rise in vegetarianism is largely down to the effects of Peter Singer's ideas and logic permeating into every layer of society.

His importance is paramount: with his book *Animal Liberation* (1975) he single-handedly brought the animal rights movement to life and gave ethical vegetarianism and veganism an impetus in the Western world which it hasn't lost since. The evidence is in the supermarket shelves and the 'vegan-friendly' labelling of all kinds of products.

Animal liberation: pain-centred

Singer is hailed as the 'founding father of animal rights' and says of himself that he is a 'consequentialist'. The gist of that persuasion is that right and wrong are determined by the consequences of an action and nothing but the consequences. There is no moral right or wrong as such, it's the consequences which make actions right or wrong. It is important to note that for a consequentialist, legally right is not morally right, because legal rights do not necessarily generate the best consequences. Singer thus does not believe in legal or moral rights but only in consequences and uses 'right' and 'animal rights' as 'a convenient political shorthand'.

The foundation of *Animal Liberation* is straightforward:

1. Whatever does not feel pain or suffer has no interests which we need to consider

2. Whatever feels pain or suffers does have interests which we must consider

3. Equal interests means equal consideration and the morally right course is what produces the 'best consequences for all of those who are affected by our actions'

This is fairly open-ended because the 'best consequences' could be any outcome depending on how you load the dice. Recreational fishing could be perfectly OK depending on which way you look at it. According to Singer and his acolytes, auxiliary concepts (*see below*) like 'unnecessary suffering', 'benefit of doubt' and 'precautionary principle' mean that recreational fishing and *all* fish use should be banned. From the point of view of pain-centred philosophers and activists, the FFP question is about three interlinked areas:

1. A better cruelty-free, PC, all-inclusive world
2. Getting rid of recreational fishing
3. Fish

Fish are instrumental in disposing of an undesirable species: the fisher. The reason for this, they hold, is the anglers' immature mindset and intellect which fails to understand the rights of animals, in this case fish. Anglers like hunters are obstacles on the march to a better and cruelty-free, just, diverse and all inclusive etc. world. Recreational fishing and hunting are reactionary bastions of unreason which have to be stormed in order to advance moral progress. Progress, the FFP camp firmly believes, goes hand in hand with ethical vegetarianism and veganism.

Peter Singer and consequentialists all over the world hold that where there is no pain there is no moral concern because no pain means no interests which we need to take into account. What does this mean in practical terms?

Fishing first: the presumed pain in fish means that Rudy Rainbow and friends have interests which we have to consider. On balance the interests of the fish (pain) outweigh the pleasure of the angler, therefore angling is morally wrong. Note the wording: this is not about cruelty but about accounting for or disregarding interests – the purity of the doctrine. Of course cruelty is often brought into it to buttress the case against the angler and fishing but the consequentialist essence is that pain = interests. And should there be any doubt, there is always the recourse to 'unnecessary suffering' which ultimately boils down to 'because I say so'.

To drive home this important point, consider the following. For the sake of this argument, we will assume that insects feel no pain (though that is a whole other story for another time). There are philosophers who advocate we should farm and eat insects because there is no pain involved making it morally acceptable food for human beings. Consider now a beautiful butterfly: you could catch one and tear off one wing after the other. Same with the feelers and eyes. You could maltreat and dismember it any way you like.

All of that is of no moral concern because no pain = no interests.

This merits a closer look:

1. Pain-centred philosophers and activists have no issue with dismembering the butterfly. No pain, no problem. It is absolutely consistent with pain-centred logic.

   Note: So, according to this philosophy, to undertake cruel and revolting acts on humans or animals is by itself not a moral wrong: it's the consequences that make an action 'right' or 'wrong'.

2. Most people who are not animal rights philosophers and activists would use their common sense and moral intuition to judge the wanton act of dismembering the butterfly as plain wrong.

To this the pain-centred philosophers can only give one answer (as Singer does): *there is no such thing as plain wrong!* It's about pain, interests and consequences. The butterfly doesn't suffer pain therefore it has no interests we need to take into consideration. Since it has no self-awareness or self-identity it has no rights either (*see next chapter*). Should they try and wriggle out of the butterfly dilemma by taking recourse to, for example, 'respect' or 'common sense' then they accept that there are values which matter and not just consequences. This then nullifies and trashes their entire case which is that if there is no pain, there are no interests.

There are other quite counter-intuitive conclusions that highly educated, highly paid and highly respected university teachers reach. One particularly repugnant but logical one is that training guide dogs for the blind is morally wrong because it is equivalent to the enslavement of black Africans. In this line of thinking, there is of course no room for police or search dogs nor companion animals generally.

To his credit Peter Singer is ruthless in following his logic wherever it might lead him. While fishing is a strict no-no, sex with animals, children or corpses is absolutely no moral problem provided that animals, children and corpses don't feel pain or suffer in the act.

Anybody wishing to understand Singer and his philosophy should watch William Crawley's interview with Peter Singer. It's all there in a nutshell in part three – the section on paedophilia. See Uncle Google. There is no better summary of the man and his philosophy.

To be scrupulously fair: Peter Singer doesn't advocate the said practices, he merely says they present no moral problem. Singer's logic might lead him anywhere. If, for example, vivisection would produce 'the best consequences' for all concerned, then vivisection would become perfectly OK for Singer. As a matter of fact he has said as much: 'It is clear at least some animal research does have benefits'.

On the face of it this seems astounding but there is consequentialism for you. To all intents and purposes, recreational fishing might produce the 'best consequences' but that, of course, would require loading the dice differently. That's it: pain-centred animal liberation philosopher Singer is basically easy to understand. There are no mysteries. The potentially confusing fact is that Singer doesn't believe in rights but uses the term as a 'convenient political shorthand'.

Animal rights: rights-centred

The second man who stands between you and your fish – if you are an angler – is Tom Regan. He too is a philosopher but less famous and of a different persuasion: he is a non-consequentialist and firmly believes in moral rights. Neither consequences nor pain are of any real significance. It's rights, rights and rights all the way. Regan's major contribution *The Case for Animal Rights* was published in 1983 and in essence it claimed:

1. All animals that have a sense of identity are 'subjects of a life'.
2. All subjects of a life have inherent value and must be respected. Inherent value does not come in degrees; all subjects of a life have equal inherent value. The most basic respect is to respect their integrity. Human beings have no moral right to use animals i.e. subjects of a life have an unconditional right to life.

Of fish and fishing Regan says that fish might not fulfil the criteria for a subject of a life but we should nevertheless not fish because that encourages a hostile attitude to the inherent value of animals that *are* a subject of a life.

So even if an animal is not a subject of a life we should not in any way interfere with it because indirectly this might lead to harm to a separate subject of a life. Implicitly and in practice this means hands off of all animals. While Peter Singer theoretically and in practice leaves all doors open, Tom Regan slams them all shut: all animal use by human beings, including recreational fishing must end. It is perfectly clear.

Animal liberation, animal rights and animal welfare are three distinct, fundamentally different concepts. However, it is important to note that in 999 out of a 1,000 cases what is meant by 'animal rights' is either Peter Singer's pain-centred, consequentialist theory or a modified, extended, softened down or embellished version thereof. Many thousands of authors, teachers, lecturers, professors, journalists, science activists, lawyers, politicians and NGO executives surf on the pain-centred, consequentialist wave. I use 'animal rights' to cover both views: consequentialist (Singer) and animal rights view (Regan) which is in line with most general usage.

Animal protection and animal welfare

There is a third 'philosophy' with no philosopher's name to it which is 'animal welfare'. Modern animal welfare is largely pain-centred too but it regards the use of animals by humans as permissible in principle provided certain welfare conditions are met.

Note: animal welfare started out as 'animal protection' (from cruelty) and, for example, has only been called animal welfare in England since 2007. That is a significant shift from a passive principle (protection) to an active one (welfare). In order to promote welfare you have to know what welfare means which is one of the reasons why in recent years animal welfare science has proliferated.

Not so long ago the 'certain welfare conditions' to be met were couched in very loose terms. Animal welfare science today reached a point where rule of thumb is out of the question. This applies to the definition of 'animal welfare' itself. The American Veterinary Medical Association defines animal welfare as follows:

> An animal is in a good state of welfare if (as indicated by scientific evidence) it is healthy, comfortable, well-nourished, safe, able to express innate behavior, and if it is not suffering from unpleasant states such as pain, fear, and distress.

The European Union's five freedoms echo exactly that:
- Freedom from hunger and thirst
- Freedom from discomfort
- Freedom from pain, injury and disease
- Freedom to express normal behaviour
- Freedom from fear and distress

This can be taken as representative of most First World concepts concerning animals in human care but obviously can't apply to animals living in the wild because all of the mentioned negative experiences (hunger, thirst, predation etc.) occur naturally out there in the wild. However if all animals are equal, as often claimed, there is an injustice here. This can of course easily be remedied by ending all animal farming and animal use *(see Manifesto page 124)*.

What does animal welfare mean for fish and recreational fishing? All fishing causes some stress, and if you believe the FFP faithful, pain and suffering, therefore fishing creates a welfare issue. If we assumed that fish did not feel pain, would that mean there is no welfare issue? From an animal rights point of view, there then would be no welfare issue, and you would have moral licence to maim and mutilate as many fishes as you like. Nobody in their sane mind does this because, as in the earlier example of the butterfly *(see page 64)*, it's plain wrong.

The question remains: do fish receive welfare if they feel no pain? The answer for a variety of reasons is, of course, an emphatic yes. 'Pain' is a cheap ideology-driven tunnel-vision in relation to fish, not least because it blinkers us to the entire environment on which they depend and hopefully thrive. Fish welfare can, for example, be compromised by pollution and the question of pain doesn't come into it at all because the fish is but one of many organisms that 'suffer' in an ecosystem crisis.

A telling example of this is the issue of endocrine disruption, which is known to occur when fish populations are exposed to synthetic oestrogen (think 'the pill' or hormone replacement therapy drugs flushed into the environment via sewers) or other oestrogenic chemical pollutants. It has been proven on the 'whole of lake' scale in Canada, that these pollutants can result in the collapse of entire populations of fathead minnows, an important baitfish. Exposure to extremely low concentrations of these chemicals does not 'cause pain' to the fathead minnows in any conventional sense, yet their entire population will collapse once the male fish in the population become feminised to the point that they can no longer fertilise the female eggs.

Such examples demonstrate that fish welfare should be embedded in a wider environmental context. And given that fish themselves must exist in food chains where predation on other fishes is absolutely necessary in order for them to obtain their own 'five freedoms' it is surely not far off the mark to say that the best basic welfare provision for fish is a healthy aquatic environment.

Indeed Australian-based aquatic animal health specialist Dr Ben Diggles has cast this into the fish welfare formula:

No habitat / poor water quality = no fish, and no fish = no fish welfare

Habitat and water quality = fish welfare.

Pursuing this line of thought leads to: what's good for fish is, as a rule, good for humans.

What does fish welfare mean in the narrower picture of practical recreational fishing – assuming there is no pain? Some examples of welfare are:

- proper handling in catch-and-release
- immediate killing of the fish caught
- barbless hooks
- minimal tackle requirements (e.g. line strength, hook size)
- quotas/bag limits
- minimal size keepnets or none
- other selective restrictions or prescriptions (like killing invasive species if caught).

A pragmatic fish welfare approach is focused on what actually happens to the fish and its environment. It tries to improve fish welfare in recreational fishing by looking at the objective impacts and outcomes, namely health, fitness, survival rate and, not least, the environmental impact of recreational fishing and other activities.

A feelings-based approach, on the other hand, fishes in the murky waters of mental states of fish, something about which we know precisely nothing.

Under the traditional flag of animal welfare, recreational fishing is permissible as long as welfare concerns are met. However that flag may be false if animal rights pirates seize the ship and take a different course. In every tackle shop in the UK there used to be a donation tin for the RSPCA – they're gone now. In the 20 years since 2003 none of the previously outlined broad philosophical and welfare ideas and basic positions have changed but for marketing and political reasons they have hybridised to a bewildering complexity and have found their way into legislation.

There is a subtle process underway, transforming animal protection into animal rights (legal rights) via the 'back door' of animal welfare. This process runs under the heading 'animal mainstreaming' and follows the same idea as 'gender mainstreaming'. Mainstreaming an issue is to make it a policy concern. The UN 'Agenda 2030 for Sustainable Development' states under paragraph 9:

> We envisage a world in which every country enjoys sustained, inclusive and sustainable economic growth and decent work for all [...] One in which humanity lives in harmony with nature and in which wildlife and other living species are protected.

That sounds harmless enough but:

> Mainstreaming animal welfare concerns into sustainable development policy will require transformative changes to key industries, practices and values, and it may encounter resistance from interest groups.

Recreational fishing will be one of them.

Over the years since 2003, the FFP-science has mushroomed and reached mainstream status. The same research ideas and methods and the same animal rights tactics have recently been applied to crustaceans. The Crustacean Compassion organisation reported recently:

> For us, 2022 will be best remembered as the year that the Animal Welfare (Sentience) Act became law and for good reason, as this was the very first time that animals like crabs, lobsters and prawns (decapod crustaceans) were included in UK legislation. Prior to this, they had no more protection than a Brussels sprout.

That's very unfair to the Brussels sprout because it is sentient too (*see below*). And lest you think this sentience business is nonsense, consider the changes in animal welfare laws in the last 20 odd years. 'Sentience' found its way into the EU constitution in 2007; New Zealand (2015), France (2015), Portugal (2016), Columbia (2017), Spain and Britain (2022) followed. Switzerland (1992) was and still is ahead of everybody else: 'dignity of animals' has its place in the constitution. This is animal rights and its underlying philosophy cemented in law!

Paradise now!

The OED online defines 'sentient' as 'to see and feel through the senses'. 'Sentience' is an evolving concept and a playground for philosophers and lawyers alike. For many it means 'capable of feeling pain' but the scope of sentience can also include 'positive and negative emotional states', 'expectations about the future' and other states or experiences which matter to the individual animal or plant. There are so many definitions out there that you can't see the wood for the trees. I take the OED as an anchor point because all life feels through the senses. However not everything that 'senses' is alive. For example, a smoke detector. Unlike organisms it doesn't need its senses in order to survive and to reproduce. The smoke detector is not sentient but all life is and that includes the Brussels sprout, bacteria and us.

The Brussels sprout like all other plants strives to reproduce and by harvesting and eating it you're actively preventing it from reaching its natural destiny. If you take sentience as seriously as Jainism does, you walk around with a face mask so as not to accidentally kill a tiny insect with your breathing and you eat not much more than nuts. That would be too uncomfortable for the philistine Western animal rights professors which is probably why they put up the pain barrier where it is convenient, so they can at least munch Brussels sprouts and the like. Never mind all the slugs, mice and other sentient beings which are killed in the production process of Brussels sprouts.

The Jains have a point. In pursuing its goals, all life is in its own way unique and in that sense all life is equal and of equal value. Equality reigns likewise from a different angle (more on beauty later):

All things bright and beautiful,
all creatures great and small,
all things wise and wonderful:
the Lord God made them all.

Incidentally: replace 'the Lord God' by 'evolution' and then even the most ardent evolutionists can sing from the same hymn sheet. Darwin himself wrote: '…endless forms most beautiful and most wonderful have been, and are being evolved.'

Enter animal rights; exit beauty. All animals are still equal but some are more equal than others. Sentience/pain – calibrated on human pain – is eminently suitable for pushing the agenda of the cruelty-free, better, all-inclusive world because most people like, love and admire animals. Sentience is also perfect because by now its use tacitly includes pain. An achievement for animal rights.

Let's spare a thought for non-sentience: all life depends on non-sentient physical and chemical processes (e.g. rocks, oxygen, water), as explained for example in the Gaia hypothesis. Common sense would therefore suggest that we surely have a moral obligation to the non-sentient foundation of life. But no. No pain, no self-awareness, no interests. The same goes for rivers, lakes, jungles, deserts and the climate. Species protection? Conservation? Biodiversity? The agonising logic: no sentience, no moral value of any kind.

Far fetched? Far from it: the utilitarian sun king Singer states: 'Pain and suffering are in themselves bad and should be prevented or minimised, irrespective of the race, sex or species of the being that suffers' – equal suffering, equal consideration. That leads straight to preventing predation in nature. Nature is full of 'pain and suffering', full of 'bad'. It means you would have to interfere to save the mouse (a 'person') from the fox (another 'person') or the antelope from the lion. But it does not end there: it culminates in 'paradise engineering' which envisages genetically modifying predators to create, for instance, vegan lions. Screwing up entire ecosystems by preventing pain in the name of animal rights is absolutely no logical challenge to them. It would be no practical problem either if it could be done. The proponents and sympathisers of such views are respected academics, scientific advisors to humanitarian societies and members of ethics commissions dealing with – among other things – recreational fishing.

Completely bananas!

To soup up their rhetoric, animal rights philosophers and activists deploy some auxiliary concepts lest their case seem too simplistic. In the last 20 years, tons of toner and quintillions of pixels have been deployed to rewrite or paraphrase the same ideas over and over again. Quantity presumably is meant to change into quality at some stage. And it does: 'anthropocentrism', 'speciesism' and especially 'sentience' are now popular buzzwords. Their meaning depends to a great degree on who is using them and how they are used. They are a true asset in the animal rights' armoury because they can confuse even those who use them. Nevertheless they change nothing, absolutely nothing, about the basic positions I have outlined.

Anthropocentrism

Literally translated, anthropocentrism means human-centredness. The OED online dictionary defines the core of anthropocentrism as the 'primary or exclusive focus on humanity; the view or belief that humanity is the central or most important element of existence…'

Animal rights philosophers assert that anthropocentrism means that only human beings have moral status and that animals are of no moral concern and of instrumental value only. Instrumental value means the moral status of animals is determined by the degree to which they are useful to us (e.g. for food or as pets).

Anthropocentrism comes in many guises but the Christian variety is usually singled out as being responsible not only for all the wrongs done to animals but also the environment. Why? The Christian view is accused of encouraging a negative attitude to nature: 'Be fruitful, multiply, fill the earth, and subdue it. Have dominion over the fish of the sea, over the birds of the sky, and over every living thing that moves on the earth.'

This is seen as the biblical endorsement for degrading animals and the environment to instrumental value only. To animal rights philosophers and supporters, anthropocentrism is, so to speak, the devil itself.

The exorcism proceeds in two steps.

Step one: human beings must abandon the idea that they are the pinnacle of creation (human supremacy by the way works also without a Christian background – there is also atheistic anthropocentrism).

Step two: human beings have to expand their range of direct moral concern to animals – or to certain animals. A frequently used visualisation of this is the expanding moral circle.

This is how it works: in the traditional view, so it is claimed, only humans had moral value (our species). With animal rights the circle expands to mammals and then to all sentient beings (pain, interests etc.). The circle is expanded not because any mammal or sentient being has asked to be included in the human system of moral values but because animal rights philosophers (human beings) bestow moral value on them. It is a human choice.

The expanding moral circle itself is a uniquely human creation. Humans are at the centre and in control of the moral realm. The expanding moral concern and circle is human-centred and an anthropocentric construct. That is hardly surprising because like all other species we humans are centred on ourselves. We can't help being human just as seals can't help being seals and as such are seal-centred.

This particular moral circle is redrawn from a PC and animal rights aligned website and is representative of thousands of its kind. All of them end their expansion with sentient beings. The reality

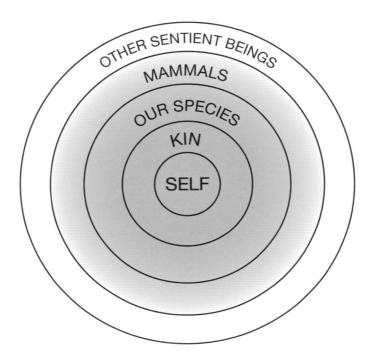

outside sentience i.e. the earth and its ecosystems are – in the animal rights view – morally non-existent because they don't feel pain and therefore have no interests which we must morally consider. Animal rights philosophy is, so to speak, caged in by its own circles: they can't think or act outside their narrow boundaries.

Within those boundaries they relentlessly attack especially Christian anthropocentrism as the source of all evil that befalls helpless animals.

However, the more animal rights philosophers pontificate about evil anthropocentrism, the more anthropocentric they get. When they accuse others of anthropocentrism, they inadvertently incriminate themselves because, like all other human beings, they are by their very nature centred on themselves.

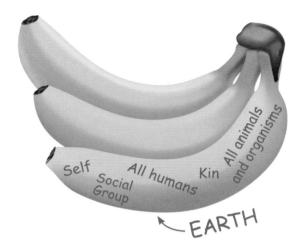

An alternative outlook which can accommodate almost any kind of expansion is the moral banana. The banana has absolutely no connection (unlike an apple would have) to the much despised Christian tradition, without excluding it.

All life on the banana is equal because we are all sharing the same banana. All life is of moral value (not just life that feels pain) and ecosystems and the earth are of moral value. Unlike in the moral circle, with the banana there are no rigid geometrically defined boundaries and neat compartments. The banana view implies that all life and its physical basis (earth) must be on our moral radar because no banana, no life. How can we not include the basis of life in our moral considerations? Practical and moral reflections on the nature of life open the door to environmentalism and conservation.

Speciesism

'Speciesism' is a monstrous concept coined by Richard Ryder:

> The word speciesism came to me while I was lying in a bath in Oxford some 35 years ago. It was like racism or sexism a

> prejudice based upon morally irrelevant physical differences. Since Darwin we have known we are human animals related to all the other animals through evolution; how, then, can we justify our almost total oppression of all the other species? All animal species can suffer pain and distress. Animals scream and writhe like us; their nervous systems are similar and contain the same biochemicals that we know are associated with the experience of pain in ourselves.

Note the link to 'racism' and 'sexism'. Something important happens here: the animal rights cause is linked to social causes concerning human beings. However it is not only linked, but put on the same footing. Prioritising say 'sexism' because it concerns human beings over animal rights would already be speciesist. Furthermore, equating speciesism with 'racism' and 'sexism' makes my fellow men my fellow animals!

I wonder whether Ryder's theory of 'speciesism' would have achieved the same amazing trajectory had it been discovered in a bath or cold shower in say Grimsby. However, Oxford it was: Singer and other animal rights philosophers and activists the world over have taken the concept on board, popularised it, refined and tweaked it to their needs.

Singer's version of 'speciesism' is probably the most quoted and the one I go by:

> a prejudice or attitude of bias in favour of the interests of members of one's own species and against those of members of other species.

All animals are speciesist. Their prejudice or bias couldn't be more so. It's absolute. The lion's bias is the same immutable absolute as is the tardigrade's (microscopic creature also called 'water bear'), the squirrel's, the trout's or the seagull's. If animals weren't biased in the speciesist sense they wouldn't survive either as individuals or species. It would be patently absurd to accuse animals of speciesism except, of course, the 'human animal' which is asked to act against its own interests. No animal, except the 'human animal' is capable of understanding this and acting accordingly. So the human animal has been singled out from other animals as different... That is pure anthropocentric! Ryder has jumped from the frying pan into the fire and there is too little water in Ryder's 'Oxford' bath tub to extinguish it.

Sentience and personhood

We take our cue again from Peter Singer because as in other areas he has defined the playing field. In a letter to *The New York Times* he writes:

> ...First, I have never denied that newborn humans are sentient. That would be a crazy position. Obviously babies can feel pain as can non-human animals. My view is that newborns are not self-aware, that is, do not have a sense of themselves as a separate object, with a past and a future.

> ...I think that every sentient being deserves to have its interests given equal consideration. [...] I don't think that sentient creatures have a personal interest in continuing to live, unless they are also self-aware beings.

Note: sentience *per se* is not sufficient for a being to enter into the moral sphere of animal rights. There's also got to be self-awareness in the Singerian and other consequentialist cocktails in order to qualify for a place in the human moral sun. There are two scenarios here:

1. Sentience without self-awareness
2. Sentience with self-awareness

Sentience without self-awareness

All life is sentient. An organism however simple or however complex can thrive only if it's able to feel through some senses what is important for its existence and reproduction. Pain and self-awareness might not be vital to some organisms such as tardigrades and plants but they are still sentient. The import of self-awareness emerges clearly when Singer and others ponder on what the absence of it means.

Being sentient but not self-aware, the human baby can be killed (murdered?) with no moral harm done. Likewise a dementia sufferer can be 'euthanised' (murdered?) because he is presumed to have no self-awareness. On the other hand a healthy chicken is sentient and self-aware and must not be killed, for food or any other reason and we must not steal its eggs. I was present at the University of Bern (Switzerland, 2008) when Singer argued for the death of the Alzheimer's patient and the chicken's right to live. The audience swallowed it without demur.

This key area merits a closer look at the reasoning which leads to such conclusions:

1. Traditionally a newborn human child is a human being.

2. From an animal rights perspective, the newborn human child is a non-person because it has no self-awareness.

3. The human baby becomes a person only when it has an awareness of itself and has developed other typical characteristics of its species.

4. Animals which have self-awareness and developed the typical characteristics of their species are persons.

5. A healthy chicken is a person but a human baby is not a person.

6. The healthy chicken on the strength of its personhood has a right to live; the human baby as a non-person has no such right.

With a few artful rhetoric pirouettes, fish too can be assigned personhood. It's easy to apply. Another popular and to my mind particularly sick line of argument is the one regarding 'marginal cases'. It's a spin-off of the self-awareness criteria above and the bare bones look like this:

1. Human beings are said to have certain 'morally relevant' defining characteristics like self-awareness, being rational, having language etc. which is why they have direct moral consideration or status.

2. In marginal cases such as people with intellectual disabilities or dementia, the 'morally relevant' defining characteristics – especially self-awareness – are presumed not present, yet they still have moral and legal status.

3. Animals lack some or all of the defining 'morally relevant' characteristics of the marginal cases too and should therefore have the same moral and legal status as human beings.

Animals thus are imperfect human beings and human beings are perfect animals. On that very basis a philosopher has recently argued that some animals should get the right to vote. There are numerous problems with this kind of reasoning. Even the name of the argument 'marginal cases' stinks. Surely rather than being classified as 'marginal', the people who suffer from a disability are human beings – human beings who have been given extra challenges by nature or fate. But they are still human beings which is why they have the same direct moral standing as human beings without disabilities.

The 'morally relevant characteristics' are another major difficulty: what exactly are the relevant characteristics? Why are they relevant? And who decides that they are relevant? The Ministry of Universal Truth? Picking out those presumed morally relevant characteristics from 'normal' human beings and then hinging the case for animal rights on 'not normal' human beings seems weak reasoning. Just imagine for a moment if there were no 'marginal cases' – what then?

The most absurd of all angles on 'marginal cases' is that it assumes that animals are a kind of imperfect human being. Can you get any more anthropocentric than this? If rights for animals are granted, the minimal respect to human beings and to animals would be to grant those rights to them as animals qua animals and not as imperfect human beings. This argument smacks of ableism and condescension.

Sentience with self-awareness

The gold standard of sentience with self-awareness is the adult human being who is a full person within their species. The nearest persons of other species are primates. The animal rights philosophers and lawyers say primates should also get the same basic legal rights as the human animal, namely human rights.

There are heaps of problems with this and I only touch on one of them for reasons of space and the reader's patience. Primates are said to be genetically 99% like us and that alone should be sufficient to grant them human rights. However, if you look at this closely, that 1% is a mighty 1%! Bonobos don't fly to the moon nor do they ponder on the meaning of life, they don't cook and nor do they tell jokes etc. Neither do nematodes. Nematodes are small worms and apparently they are genetically 75% like us. Does that mean basic human rights for nematodes too? After all, 75% human is quite a lot. The really interesting bits are not the 99% but that 1%. That 1% makes us live in another world. Anatomic and genetic similarity are obviously two different kettles of fish.

Again the self-awareness angle plays into this: I guess a worm would be classified as not self-aware (not yet) and thus of no moral importance. However worms seem to 'know' what they are doing. Instead of explaining that as preprogrammed, innate 'instinctive' behaviour, it could easily be construed as human-like intentional decision-making. To complete the survey of the main auxiliary concepts let's have a look at abilities, similarity, and necessity.

Abilities and learning

Be it lions, salmon or bar-tailed godwits, a closer look at any animal will reveal amazing feats and abilities. All surpass what I ever will be capable of achieving. I will never fly and navigate like a bar-tailed godwit nor survive extreme heat and radiation as tardigrades do. Likewise the tardigrade will never fly like the bar-tailed godwit nor will it try to make sense of animal rights like I do. Innate abilities and instincts rarely lead animals astray: animals are perfect in what they do.

Human beings, on the other hand, with their rationality and reason and their specific abilities are rarely perfect. Nevertheless in all matters 'human' I excel over all other species. Despite my human limitations, no other animal species – or rather individual thereof – can hold a candle to me.

Ravens make plans, primates learn Ameslan (do they really?), cockatoos push bricks off wheelie bins, some fish use tools: there are all sorts of clever achievements by some animals either learned by themselves or through being conditioned to perform certain actions. Animal rights philosophers and activists point to the learning abilities of some animals so they can say, 'They are like us'. 'Like us' is as popular among professional animal rights activists (philosophers and scientists) as it is with grass roots animal rights and animal welfare supporters. Indeed it is very human to look at fellow creatures, especially furry ones like cats and dogs, as being 'like us'. We feel empathetic and inclined to help them if they suffer. Who can resist the appeal of a cute dog or kitten?

Salmon par and other 'baby fish' hardly trigger the same kind of emotional response which may or may not be the reason why in the West most people don't eat dogs but have no hesitation to eat fish. Matthew Chalmers on the Sentient Media website:

> For many people, the mass killing of fish [commercial fishing] has always seemed more morally justifiable than that of other animals, and pescetarianism is a relatively common diet. This is likely because fish look and act so different to humans that they seem too strange — too distinct and distant — to be considered in any way *like us.* [italics added]

Like the genetic similarity above 'like us' needs to be put in perspective. However cute or clever an animal may be, all animals are still light years away from human abilities just as we are light years away from the achievements of animals. We're different. If ever there was a bogus claim it is that fish or other animals are 'like us'. True, they have evolved from the same common beginnings and are thus related to us. However, nature has not shaped all creatures the same. Some are very, very different.

Consider the difference between a tardigrade and a cockatoo: although both are created by the same Mother Nature and both are sentient relative to their environment they couldn't be more dissimilar. It can't be emphasised enough we are worlds apart even from our nearest cousins the primates. They and all other animals are so manifestly not 'like us' that I really wonder how people get away with the claim that 'they are like us'.

The truly remarkable part of this matter of abilities and similarities is that animal rights philosophers and activists measure animal behaviour by the human benchmark. What could be more anthropocentric and speciesist than to carry the human yardstick into the animal world and then attribute moral value to animal species that seem to come close to the human species (sentience, pain, self-awareness). And what of the poor animals which are not like us? A rightless caste? Humankind is putting itself up as the measure for all things! The more things change, the more they stay the same.

A more sober angle looking at abilities and similarities is perhaps that what unites all animals and human beings is that they are all radically different despite a common origin. If the sequence of big bang – primeval soup – evolution is anything to go by, then genetic similarities between species are hardly surprising. However, that kind of similarity doesn't warrant conclusions about similar abilities or experiences. Visual similarities likewise are not reliable guides to identify experiences, states or facts: what looks like a fish might turn out to be mammal and though I look relatively similar to Einstein, that's as far as the similarity goes.

Necessity

Probably the most popular argument for animal rights in regard to recreational fishing is 'unnecessary suffering'. As one social scientist put it:

> The interest that anglers demonstrate in sport fishing is recreational and not a basic, or necessary, survival interest.

'Necessity' is the perfect knockout argument for literally everything in any context because everything beyond bare survival need can be dismantled at will. Take, for example, all those climate summits: was it really necessary that all those Hollywood stars and politicians flew in their private jets to Rio, Paris, Sicily, Glasgow, Sharm El Sheikh and all those other beautiful places? No, they caused unnecessary carbon emissions, and we know we should all be avoiding unnecessary carbon emissions. Was it necessary that the participants met in person? No, there were alternatives where they would still have reached same result. Take television sets, cars or fishing boats, works of art, pets, sports – everything can be argued back to the Stone Age – even animal rights. Even civilisation and culture are unnecessary achievements.

Are reading, writing or maths 'necessary'? No, mankind in its earlier stages and all the animals on the planet have survived without it. Apart from a few basic activities like breathing, hardly anything we deem 'necessary' is necessary. Should for some reason the argument about necessity reach an impasse, the easy way out is to give the benefit of the doubt. The very last resort is the precautionary principle. In the context of recreational fishing the necessity-cum-precautionary principle argument would run as follows:

Recreational fishing causes 'unnecessary suffering'. You don't need to fish; it is not a necessity. And even if there are doubts that Rudy Rainbow suffers, we will give him the benefit of doubt and out of an abundance of caution (precautionary principle) you should stop fishing.

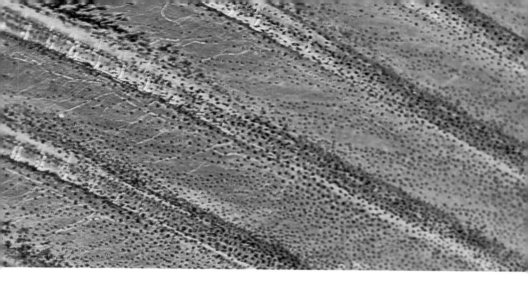

Chapter 3
Breaking Down Barriers Between Humans And Animals

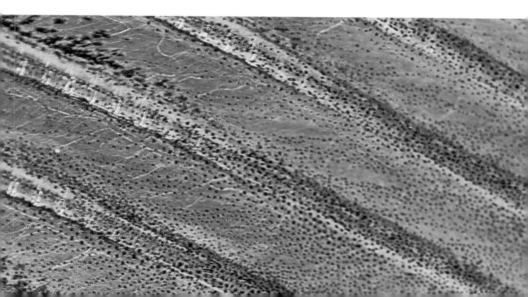

Animal rights philosophers spare no effort to persuade the world that we humans are just one sort among other animals. If as a species and individuals we put our interests first, that is morally wrong. We should stop thinking in hierarchical terms, pull the species barrier down and treat animals as equals. This is why personhood, similarities, abilities, learning etc. are so important: anything that levels us up with cockroaches (sentient, highly intelligent, able to learn) is morally desirable. I am going to pick out two points of the animal rights movement's multi-pronged attack on the human being: the species barrier and language.

The **species barrier** in the moral sense is said to be a major obstacle on the way to a better world which is why we should pull it down and grant animals the same moral and legal rights. This is also labelled as 'extending personhood' to animals. Whatever you call it, the idea of breaking down the species barrier is unilateral. No animal in its sane mind would contemplate extending its idea of personhood to humans nor would it accept our idea of personhood. All animals innately know and 'insist' on barriers – the barriers are natural and there for a reason (including preventing diseases transmitted from one species to another). Lions know what their 'personhood' is and insist on it and behave accordingly and so do all other species. Animal rights wants us human beings to head in the opposite direction.

I already touched on **language** briefly but as this is an important area a few additional remarks are warranted. According to the animal rights view, commonly used words such as human, man, mankind are highly suspect. Even woman*kind (the star signals the transgender inclusion) is suspect. Using words like 'human' or 'people' cements the special status of human beings in nature and that is undesirable and wrong. Instead we should learn to use words like 'person' or 'animal' when we mean human beings. And should we talk of animals qua animals i.e. non-human animals, we should always include ourselves.

This means if I say 'dog' or 'worm' or speak of any other animal, it will also mean me. Humanity has to be banned from language. Note: this is all focused on 'us', which is anthropocentric to an unacceptable degree and, worse, of no beneficial consequence to any animal. Language control (Newspeak) is more than just vaguely reminiscent of *1984*.

The trap of personhood

Human babies do not innately know what the rule of law is. They grow into it. As adults they understand what it means and explicitly or implicitly consent and comply. Animal babies do not innately know what the rule of law is either but they never grow into it and will never understand what it means. They can't in any way or at any stage of their lives consent or comply. Animals are 'trapped' in their unique being and their own 'culture' just as we are in ours. Their brains are just not 'made' for human culture and concepts. At whatever stage of their lives neither an elephant nor a mouse is a human baby and vice versa.

As pointed out above: no animal can want or not want our moral and legal rights – it is just beyond them. They haven't got the brains for it. Their brains have evolved for their needs and their 'rights' isn't one of their needs.

Sandra was born in a zoo in Germany and then was moved to various other places and finally ended up on her own in Buenos Aires. Hers is a terrible story, absolutely terrible, a disgrace to all involved and full marks for the people who took up Sandra's plight. Lawyers engaged on her behalf went to court successfully demanding personhood for her. Buenos Aires zoo was closed and Sandra is now in the care of the 'Center for Great Apes' in Florida where apparently she is well looked after and no longer in solitary confinement. By being happy about this outcome and unreservedly applauding those who brought it about I am not endorsing animal rights. Cruelty is wrong, regardless of animal rights, and if I understand Sandra's story correctly it was cruel.

As I said, human culture evolves and animal rights has caught on. In what is described as a 'landmark decision' an Argentinian judge, Elena Liberatori, ruled that the orang-utan Sandra is legally not an animal but a non-human person (*persona no humana*). The judge is reported to have commented: 'With that ruling I wanted to tell society something new, that animals are sentient beings and that the first right they have is our obligation to respect them.'

Now that is a tall order. In its sweeping generality it says all and nothing (assuming the translation is correct). 'Animals are sentient beings'? My guess is the judge meant 'all animals like us' (vertebrates) and she didn't for example include sentient parasites or pests. Birds and other wildlife hardly respect their own parasites and try hard to get rid of them. Cockroaches are highly intelligent and sentient yet I don't see any reason to respect them. On another scale possums – without doubt non-human persons – are a pest here in New Zealand and I don't see any obligation for respect either. Having said that it would be plain wrong to subject them or the cockroaches to wanton violence or maltreatment (*see butterfly page 64*) – their demise has to be brought about as quickly and efficiently possible.

It is more intriguing to ask what a non-human person is. The concept of 'person' is a species-specific 'human' reality. It is by its nature anthropocentric, self-referencing and assigning it to all animals 'humanises' them in a meaningless way. There is no such thing as a non-human person. There are definitely non-human personalities and characters – dogs, horses, even fish (stubbornly not taking your fly/bait) have character and personality. Nevertheless 'person' qua 'person' is exclusively human. Certainly there are non-human beings, creatures, organisms or whatever but no non-human persons.

Other species do not have a concept of 'person' let alone of a non-human person. If a tardigrade did perceive itself as something like a person then its frame of reference would be its tardigrade world and there is no way we can know what that is. It's the same cul-de-sac as 'a kind of pain'. Their self-perception as for any species is as the beings they are. The concept of non-human person insults the autonomy and integrity of their being – not that they mind too much. A cockroach couldn't care less if you insult it with: 'You damn non-human person!' Whereas most people would understand perfectly the insult: 'You damn cockroach!' The answer to the question of what a non-human person is, is obvious: 'We don't know.'

Animals do not perceive themselves as cats, dogs or elephants either. That is our human projection too. What does this mean?

Do we know what it's like to be a cat? *No.*

Does the cat know it is a 'cat'? *No.*

Does the cat know it is a non-human person? *No.*

Does the cat know when it's treated badly? *Yes.*

Do we know when a cat is treated badly? *As a rule, yes.*

Does the concept of non-human person help a maltreated cat? *No.*

Does respect help a maltreated cat? *No.*

Does common sense, everyday morality help a maltreated cat? *Yes.*

'Extending personhood' adds a completely futile element to animal welfare. There is no benefit in it for either animals or us, although some of us may find it bolsters our self-esteem in occupying the moral high ground. The humane treatment of animals – all animals – requires neither sentience nor respect nor rights (extended personhood). Animals are better off protected by common sense, middle-of-the-road morality and straightforward pragmatic human animal welfare/ protection laws informed by values, needs, science, reason and especially in the case of fish – a view of the entire picture.

I am perfectly happy to be an ape

The human/animal divide is real just as the tardigrade/cockatoo divide is real. However, the power of dogma glosses over the plain fact that there is a huge divide between humans and animals: language, jokes, football, reason, cathedrals, introspection, empathy, poetry, science, the ability to grant animal rights, illusions, Diet Coke, ethics, wars, God… you name it. If you really believe there is no human/animal divide, you might also believe in a pretzel-shaped planet. After all, who says that human perception is objective: the planet we believe to be spherical may well be in reality a distorted pretzel.

Having said that, I am perfectly happy to be the Great Ape that Darwin would have me. And if I have my animal rights, the ape that I am can make choices, I can rant against animal rights, I can go fishing, I can contradict myself, I can do whatever is within reach of my capabilities but most importantly I can decide that I am human. I can also opt to try and act like a decent human and care, for example, for animals.

One of the canonic laws of animal rights is that the species barrier should disappear. Whenever the barrier is dismantled bit by bit, animal rights philosophers simultaneously build up a new one in the animal kingdom under the term 'drawing lines'. This line is to determine which animals are 'like us' (pain, self-awareness etc.) and who can enjoy the protection of animal rights; while all others, the majority of animals, would be without any rights or protection. Once again: the doctrine of animal rights excels in anthropocentrism. I think a more enlightened view would be to say that all life wants to live and succeed and that there is no divide across life, absolutely none. The divides are between the species. All species have their own way of dealing with what nature has handed out to them.

Animal rights are conservation wrongs

'Animal rights' does not live up to its name. As we've discussed so far, animal rights means rights for those animals we deem 'like us', but excludes not only the majority of animals but all organic life and the non-organic basis of life. In other words: by its own logic it has to ignore nature as a whole. Nature, the environment, any species that is not a 'person' can have no rights because it is not an 'animal like us'. This makes nonsense of, for example, species protection, fighting climate change and promoting biodiversity.

Animal rights are not compatible with conservation concerns. No issue involving an invasive species threatening another species with extinction can ever be meaningfully tackled. There is no way of doing this without violating individual animals' rights. Take the case of the kea. The kea is a beautiful, clever and charming native New Zealand parrot. In order to protect it, invasive species like the possum or the stoat have to be, ideally, exterminated. There are no two ways about it: it is either the indigenous kea or the invasive species who carry the day. There is no room for individual animals' rights here.

Likewise there is no way ecosystems can meaningfully be protected under the heading of animal rights – not in theory and not in practice. The possum (introduced from Australia) in New Zealand is described as an 'ecological nightmare'. Estimates of their numbers vary considerably but the general agreement is that there are tens of millions of them. Fighting this threat to native forests and other animals means trapping, shooting or poisoning the possums – if they can't be got rid off then their numbers must be kept as low as possible.

Possums are sentient beings and their most basic right is the right to live. 'Let justice be done, though the world perish.' Some animal rights philosophers try to gloss over this yawning abyss but others have no hesitation in sticking to their guns and following their logic to the bitter end. Tom Regan for one says that wildlife management should focus on leaving animals to their own devices – regardless of the outcome. Hats off for consistency. On the other hand, if an animal rights philosopher says – in whatever roundabout way – that the environment comes first, then he has destroyed his own argument. However, if it's expedient and for the cause of a better world, then inconsistency doesn't really matter: you can borrow from mutually exclusive concepts like animal rights and environmentalism and fudge them into, say, 'compassionate conservation'. As I discuss later, the same rabbit is pulled out of the hat when it comes to legal reasoning.

Anything goes – but let's look at Compassionate Conservation first:

Compassionate Conservation

> Compassionate Conservation is based on the ethical position that actions taken to protect biodiversity should be guided by compassion for all sentient beings.

The Hebridean hedgehog saga is probably a good instance of 'compassionate conservation'. Hedgehogs were introduced into the Hebrides in order to get rid of garden slugs. They proliferated and found even tastier food: birds' eggs. Hedgehogs threatened rare wading birds and a cull was proposed. Here is what happened [Scottish Natural Heritage]:

> However the cull attracted widespread criticism operation [sic], and it was later established that the animals did not have to be killed but could be successfully relocated to the mainland, and that work has continued. 'Since its start in 2001 to present [2017], £2,679,362 has been invested and 2,441 hedgehogs removed. This is a significant success story,' the agency said.

Compassionate Conservation holds that sentient beings are 'persons'. 'How is one to act ethically if every act holds the potential to harm fellow persons?' the compassionate vanguard asks itself. The answer?

> There is no easy answer... if one takes seriously the notion that all sentient beings are persons, forming and pursuing conservation objectives founded on mass killing would become inconceivable. The default of domination would be replaced with a default of compassion. This does not mean that one never harms a person nor that there cannot be variations in our obligations to different persons [...] Between perfectly equal moral status for all and categorical moral segregation of the few lies a wide expanse where a more inclusive and contextual moral terrain can be explored.

'Conservation objectives founded on mass killing'? I can't think of a single conservation objective founded on mass killing. Conservation is about life, not death. This bizarre moralising is plain nonsense. By the time the 'moral terrain' has been explored, the kea is extinct and the possums have munched away New Zealand's native forests. And by the way, just think of the mind-blowing cost if it were ever proposed to repatriate 50 million possums to Australia. To state the obvious (that such a proposal would be expensive nonsense) is an expression of 'human exceptionalism' which is, of course, the root of all evil. However the easy way out for me to better my moral self is to embrace 'compassionate conservation' because this helps to 'dismantle human exceptionalism'.

Unless at gunpoint: no. I prefer to remain the ape I am. As these authors seemingly don't even grasp the idea of life, why give them space here? Isn't it a waste of printing ink? The problem is not that big brother is watching you but that big academia is telling you. Twenty top academics from all over the world have stitched together the 'compassionate conservation' above (some of them also signatories of the Manifesto on *page 124*).

Aliens on our own planet

If you look at animals and fish in the wild there is a lot of 'suffering' going on. As Huntingford et al. put it (*see page 30*):

> Wild fish naturally experience a variety of adverse conditions, from attack by predators or conspecifics to starvation or exposure to poor environmental conditions. This does not make it acceptable for humans to impose such conditions on fish ...

The predator angler is the culprit: the rod being the weapon. The recreational angler is depicted as an alien intruding into a natural process. Recreational anglers – aliens on our own planet.

The crux of the matter is, of course, that humans shouldn't be allowed to impose 'adverse conditions' on fish. Why not? If nature can do no wrong and if we are part of it, we can do no wrong. You could argue further that if nature is red in tooth and claw and the human animal is part of it, then the human animal is just another predator. To the fish it is all the same whether they end up in a frying pan or the stomach of an osprey.

Animal rights authors like Huntingford et al. have no other option than to rely on our humanness. They have to appeal to the moral sense, reason, empathy and all the other features which makes humans irreducibly human. Thereby they acknowledge human exceptionalism and anthropocentrism (moral appeals make no sense to any other species nor are there moral appeals in the animal kingdom) which is exactly what animal rights activists want to get rid off. They incessantly chase their own tail and they don't seem to get tired of it.

Nature can do no wrong and yet it seems 'cruel' too: death is often slow. A cormorant-maimed fish struggles on for hours and days, most fish are gulped and digested alive (e.g. Harry Herring, *see page 54*). Eat and avoid being eaten rules in nature: and such is the case from a human point of view. Animals don't think about it too hard. Having said this: now look at domesticated and farmed animals. In comparison to their wild brethren, they live the life of Reilly (protected from predators, given health care, regular food, shelter etc.). Remember now: equal suffering = equal interests = equal consideration. This applies not just to farmed animals but also to wild animals. This means (as pointed out *page 72*) we would have to interfere with wild animals to prevent suffering and promote happiness.

Are we all aliens? If it is not acceptable for humans to impose 'adverse conditions' on fish then it surely is likewise not acceptable to impose such conditions on other animals.

There is no outdoor pursuit which does not directly or indirectly negatively impact some creature, big or small. The careless paraglider frightens wild animals into flight and potential death and the hiker crushes countless little critters with his boots. Whether or not an outdoor pursuit 'targets' animals (fish) doesn't matter because the cause of suffering (angler, glider, hiker etc.) doesn't change the suffering or death of the individual animal. The crunch is: who decides and on what grounds, what is or is not acceptable?

The easy way out would be a blanket ban on all outdoor pursuits. Since that's not on, recreational fishing would nevertheless be a good start.

There is no such thing as a cruelty-free lunch

All life thrives on the death of other life. There is suffering and death on every plate. People go vegetarian or vegan for various reasons that are not necessarily linked to animal rights:

- Health
- Quality of food
- Protest
- Climate change
- Compassion
- Lifestyle
- Religion
- Trauma

There are probably a great many vegetarians and vegans who to some extent draw on all of these angles. However ethical vegetarianism/veganism and animal rights are inseparable and that's the beastie we're interested in. There are no mysteries about ethical and militant vegetarianism: it's about preventing or reducing suffering in the animal world including, of course, fish – farmed or wild.

While the romantics in the vegetarian and vegan communities still believe their diet is 'cruelty-free' the realists freely admit that the production of vegetarian food causes pain and suffering in the animal world. Mice, rabbits, lizards and all sorts of sentient little critters are either poisoned, shot, squashed or otherwise mutilated for the cruelty-free diet. Aggravating factor: harvesting a vegetarian/vegan organic crop maims and kills more happy and sentient life than conventional agriculture does. Furthermore the environmental impact of soy bean plantations in the Amazon, for example, casts another deep shadow over the virtue-signalling diet. There is no rational way to claim that vegetarianism and veganism are cruelty-free or environmentally neutral diets.

Vegetarians and vegans are of course quick off the mark to point their fingers at the beef eaters by saying that carnivores are the real culprits, they are to blame for suffering in the animal world and, of course, for climate change. Climate change is an incidental argument for ethical vegetarianism. The last thing Elsie Shrigley and Donald Watson, the founders of The Vegan Society (1944) had in mind was climate change. The newly found society intended:

> To advocate that man's food should be derived from fruits, nuts, vegetables, grains and other wholesome nonanimal products and that it should exclude flesh, fish, fowl, eggs, honey, and animal's milk, butter, and cheese.

Over the years this evolved into a more general ethical mission statement:

> a philosophy and way of living which seeks to exclude — as far as is possible and practicable — all forms of exploitation of, and cruelty to, animals for food, clothing or any other purpose...

The nub of ethical vegetarianism and veganism is about cruelty and animal exploitation and not about climate change. Just suppose for a moment 'climate change' is stopped. That surely wouldn't turn vegetarians and vegans into carnivores.

Climate change is a godsend for vegetarianism and veganism because it gives their mission a sharp political dimension. The debate about GHG emissions from animal farming guarantees them media attention. This distracts from the real ethical question about the cruelty involved in an omnivore diet and the supposedly cruelty-free vegetarian and vegan diet.

There is no doubt that mice are highly intelligent, social and sentient animals. They sing love songs testifying to a highly evolved emotional life. The vegetarian plough mutilates them just as mercilessly as any other plough. Aside from this there are other considerations: you can't plant carrots or Brussels sprouts just anywhere. Arid land or permafrost and ice are no good for sustaining a diet rooted in virtue. In fact, due to their large size and ability to feed many with one, cattle grazing might produce infinitely less cruelty to sentient creatures than any ethically correct vegetarian or vegan soy plantation.

I can't help thinking that a little dose of biblical wisdom would do no harm: 'Why do you see the speck that is in your brother's eye, but don't consider the beam that is in your own eye?' However, as you might imagine, animal rights philosophers (ethical vegetarians and vegans) have other ideas. They are hell-bent on numbers to keep their dogmas and their superior, self-congratulating, virtuous selves alive. Just for argument sake: 100 billion sentient animals maimed and killed for omnivore food production is worse than 99 billion sentient animals maimed and killed for vegan and vegetarian food production? *Really?* The problem is, of course, live and let live is not an option for ethical vegetarians, vegans and animal rights philosophers.

'Do people need to eat what they eat?' is a popular question raised by animal rights activists. It is indeed an interesting question – if you ask it of those who want you to forego your steak or fish fillet. They could all perfectly well survive and be happy with local fruits and nuts – no problem. Their cruelty and GHG impact could be reduced to practically zero by going nuts. They don't need sophisticated vegan gourmet menus with ingredients specifically produced in the Amazon and then carted three times around the globe to please their virtuous palates.

It doesn't stop there: many ethical vegetarians and vegans keep pets which they shouldn't because it's clearly animal exploitation for human ends (entertainment, company). They actually own (how can you own an animal and believe in liberation and rights?) cats and dogs and shove vegan food down the poor animals' malnourished and

A baby tomato about to be forcibly separated from its mother and heartlessly eaten by a vegan!

STOP THIS CRUELTY NOW

vitamin-deficient throats. That is the brutal subjugation of animals in order to polish their owners' moral halos.

Of course, there's always a sophisticated way for animal rights academics to justify keeping pets, enjoying a cruelty-free diet and a lifelong university tenure. One philosopher has it that the suffering mice which are collateral damage in vegan food production are to be expected and that this is comparable to – wait for it – the annual toll of human fatalities in traffic accidents that we also accept. The life of the mouse equals the life of the human being. Equal suffering (injury, mutilation, death) equal consideration. The next thing you know there will be a demand for mouse ambulances. Incidentally, the same man says that eating meat is a sign of being weak-willed, similar to smoking. You tell that, for example, to the Maasai or the Inuit.

Moral supremacy

The ugly face of animal rights was there for all to see but nobody wanted to see it. As I described in my book *Hook, Line and Thinker* in 2003, Greenpeace, the International Fund for Animal Welfare, Sea Shepherd, PETA and do-gooders from all over the First World mercilessly pushed the northern Inuit towards poverty and despair. First world animal lovers were basically led down the garden path by the propaganda of the NGOs. Their anger and the EU's enlightened bureaucrats delivered the coup de grâce.

The Guardian reported in 2017:

> As animal rights organisations celebrated the collapse of Canada's eastcoast whitecoat sealing industry, the Inuit in northern Canada – who do not hunt seal pups, only adult harp seals – suffered from the collapse of the market for seal pelts. Despite a written exemption for Indigenous Inuit hunters, markets across the Arctic (both large-scale commercial and sustainable use) crashed.

> In 1983-85, when the ban went into effect, the average income of an Inuit seal hunter in Resolute Bay fell from Can$54,000 to $1,000. The government of the Northwest Territories estimated that nearly 18 out of 20 Inuit villages lost almost 60% of their communities' income.

> And life in these areas has not got any better since. The region is plagued with the highest unemployment rate in Canada, and the highest suicide rates in the world. A second seal ban, enforced by the European Union in 2010, only exacerbated these issues.

Not that this bothers philosophers or animal rights activists. It's business as usual along the lines of: 'There is no place for this in the modern world', along with: 'There are some universal principles of justice or preferred ways of living that we ought to defend and promote.'

The Inuit didn't stand a snowball's chance in hell to withstand the combined onslaught of animal rights in the form of NGOs and the European and other governments. Like the poor workers (slaves?) who scrape the cobalt and lithium for the batteries of electric vehicles and smart phones, the Inuit are obviously seen as expendable. People are sacrificed on the altar of animal rights for the clear consciences of politically correct Western urban vegans and the purity of their philosophical doctrine. Although Greenpeace has apologised (the other principal actors carry on undaunted) nothing has changed since 2003. In fact it's got worse. Sad and ugly.

In contrast to the International Whaling Commission which has no issue with Inuit whaling, two well-known, respected German animal rights philosophers demand that the subsistence whale hunting by the Inuit has to end. Ursula Wolf and Jens Tuider seriously suggest that the Inuit could perform a symbolic hunt for whales instead.

This is surely cultural imperialism, supreme arrogance, entitled condescension and testimony to a truly frightening mindset? Just imagine them in their cosy, well-heated studies working out how the Inuit should run their lives. It shows the true face of animal rights. Ugly.

Look on the bright side

However inhumane some of the demands and however unsavoury some of the characters involved, animal rights – for better or for worse – are here to stay. They have successfully set the arena where the gladiators of FFP science, believers and sceptics, battle it out to a cheering crowd (the media). Misanthropic to the core as animal rights may be, it doesn't mean everything coming from that corner can be dismissed out of hand. Let's look on the bright side then – yes, there is a bright side to FFP science/animal rights!

- Research into the FFP question has added to our knowledge of the biology and behaviour of certain species of fish. That in itself is a good thing.
- A positive spin-off is the research triggered into fish welfare also in the context of recreational fishing (e.g. protocols outlining best practice for releasing fish and humane dispatch of fish kept to eat); asking different questions which might lead to new insights for the good of fish. Remember: as a rule what is good for fish is good for humans.
- Equally positive is the fact that animal rights and FFP research have driven home the point that recreational fishing shouldn't take anything for granted. There is no harm in being self-critical, examining different angles and where necessary changing attitudes and practice.

However, animal rights and FFP are by no means an essential condition for the promotion and improvement of fish welfare. In fact fish pain might be the least useful compass to better the fishes lot, because we don't know what pain is for fish whereas we can know what a healthy and fit fish is. It is vital to understand that fish welfare is not a function of pain. Neither logical nor biological. Or to put it another way: there is more to fish than (presumed) pain. It is because of this that I said that pain is 'a cheap tunnel vision'.

However, assuming that fish do not feel pain doesn't let anglers or anybody else off the hook. Quite the contrary. James Rose concluded his landmark article *The Neurobehavioral Nature of Fishes and the Question of Awareness and Pain*:

> Although it is concluded from the foregoing analysis that the experiences of pain and emotional distress are not within the capacity of fishes, this conclusion in no way devalues fishes

or diminishes our responsibility for respectful and responsible stewardship. ... Our diverse uses of fishes have ancient historical precedents and modern justifications, but our increasingly deleterious impact on fishes at the population and ecological levels require us to use our best scientific knowledge and understanding to foster their health and viability.

The sanctimonious side

Peter Singer is a very successful moral entrepreneur. He makes heaps of money out of preaching how you should think, live and eat. If, for example, your mother gets dementia, then there is no point in spending money for her care (he holds that she is no longer a person and can be euthanised i.e. murdered). The money you save could be donated to an animal rights charity. That would satisfy the utilitarian calculus and produce the 'best consequences' all round.

When Singer's own mother actually got Alzheimer's, no expense was spared to look after her. When it was pointed out to him that this was a blatant contradiction to his teachings he answered:

Perhaps it is more difficult than I thought before, because it is different when it's your mother.

This is a declaration of bankruptcy. In a single sentence he pulverises his own teachings. Sensational. Not that it matters to him or his philosophical and political acolytes all over the western world: they soldier on for the better cruelty-free, just and diverse world in which, of course, there is no place for recreational fishing. All will be good if you toe the line and reel in.

A Swiss animal rights philosopher is just as blatantly hypocritical. He publicly criminalises anglers by accusing them of cruelty to animals. It's no surprise that he sits on the ethics committee which formulates recommendations to the Swiss government regarding animal welfare – fish welfare among them. All good, no worries. It is exactly the same picture elsewhere. The animal rights movement has the right people in the right places. However, this noble white knight government adviser battling for the rights of fish keeps three cats and peddles the fact to the media to signal his love for animals. During their lifetime his three cats maim, torture and kill approx. 1,000 birds and approx. 6,000 small mammals. All sentient. Note: cats don't distinguish between protected and non-protected species. Who cares? It doesn't matter. As long as other people do as they are told by the philosopher kings, all is fine.

How petty, you might say: none of us is perfect. Yes! Least of all yours truly. And you might add their real life inconsistencies don't necessarily invalidate their arguments. Nevertheless, while it is true that their lifestyle and their choices are up to them these animal rights advocates are different inasmuch as they are trying by hook or by crook to force their choices on me and you. However, if they are not bound by their own logic nor their own precepts, why should I abide by them? Because they say so, because their logical case is cogent, because they are respected university teachers, because they love cats, because they believe fish feel pain... my foot!

The dark side

For a trivial reason Arthur Schopenhauer (1788–1860) pushed an old lady down the stairs, fatally crippling her and joking about it. That was not surprising: considering his views on women, he must be one of the greatest misogynists of all time. He is also one of the high priests of 'universal compassion' for animals. Schopenhauer was the archetypal pessimist and at the core of pessimism is pain and suffering. Pain and suffering are the basic conditions of life. 'Fate is cruel and man pitiable', says Schopenhauer.

It isn't a big or illogical step to believe that human life itself is the problem. If you weren't born, you wouldn't suffer (growing up, growing old, dying). The 'best consequences' for all are perhaps not being born in the first place. Not being born is eternal bliss. 'The curse of being born' consists not only of experiencing pain but – worse – causing pain. Parents are thus seen as criminals perpetuating pain and suffering in the world – especially in the world of animals. This line of thinking cruises under the flag of 'antinatalism'.

The 'funny' thing about these people is that they are still around trying to convince us that the best thing would be that you and I disappeared off the face of the earth. After all, being dead comes fairly close to not being born. My point here again is that pain is a tunnel vision, which is why doom, gloom, misanthropy and pessimism are inherent in animal rights.

All these pain-centred ideas run to a greater or lesser degree into the same kind of problems and contradictions. The problem zones below overlap. They can't be as neatly separated as the list suggests but it helps to highlight the salient points:

- The value of a human life is relative. We are meant to 'recognise that the worth of human life varies'. All big crimes against humanity have started by relativising the humanness of humans (e.g. *Untermensch* and *lebensunwert* – not worthy of life). The best consequences can be served by all sorts of atrocities.

- Pain-centred animal rights is a completely unprincipled, value-free world view. There are no rights or wrongs. Anything goes. Animal pain *per se* is not a red line, it all depends on the consequences. Vivisection can be seen as perfectly OK.

- Animal rights is authoritarian in character because it tries to cement in law its absolutist demands based on dodgy assumptions (e.g. fish feel pain).

- Animal rights is authoritarian in character because it tries to force a way of life on you. Given free rein, diet commissars would soon be breathing down your neck.

- No animal can want or not want rights. Animal rights disregards the animals' autonomy and uniqueness as species and individuals.

- Pain-centredness is a tunnel vision: it reduces human and animal experience to one dimension. This is not just counter-intuitive but counter-factual. There is more to life than pain.

- The concept of animals being imperfect humans which therefore should be granted rights is not just flawed but it is deeply misanthropic.

- More often than not its preachers are openly hypocritical. Animal rights philosophers do not follow up their logic in practice. Most of them should end up as ascetics. Very few of them are.

- There is well-based doubt that pain is the best compass for animal welfare. A pragmatic science-based approach serves animals better.

- Animal rights divide the animal world into animals like us (sentience, pain) and all others (no sentience, no pain). Those like us get all the moral benefits. For all others, hard luck.

- If all animals are equal (those like us) we would have to interfere with wildlife in order to reduce pain and suffering there. The welfare of wild animals is the same as that for domesticated animals and pets.

- Animal rights instrumentalise animals for the movement's wider agenda of a better, meat-free, cruelty-free, groupthink world.

- Philosophically speaking it's the purity of the doctrine not the animals themselves which is at centre of the movement.

- Best consequences regardless in what form are prone to arbitrariness. It just depends how the dice are loaded and who loads them.

- While they may not suffer, guide dogs, rescue dogs and all kinds of companion animals are exploited for human ends which is seen as bad as inflicting pain. Their use has to end (*see Manifesto on page 124*).

- Conservation is incompatible with animal rights. The 'environment' feels no pain and has thus no interests which we need to take into account.

The misanthropy, pessimism and authoritarianism (think *1984*) of animal rights is the mindset of its proponents, but that is luckily not binding for anybody else – so far. The animal rights approach to recreational fishing and other fish uses pivots around the central doctrine of pain (no pain, no interests, no moral consideration). Only if fish or any other beings feel pain ('sentience') do they qualify for moral consideration.

This is why philosophers, activists and scientists with a mission are hell-bent on positive outcomes of the FFP research and why the sceptics are the proverbial red rag to them.

Without pain, the entire pain-centred animal rights philosophy collapses. Take away pain and all that is left is the wider agenda of a better world.

I can't think of any area of everyday life where 'live and let live' is an option for those subscribing to animal rights (*see Manifesto page 124*).

Winners and losers

Assuming that fish do not feel pain, you find yourself in a morally demanding situation. Unlike the simplistic FFP utilitarians you have to think outside the pain box and proactively intervene on many different interconnected levels. You have to look at the entire Mona Lisa not just her little finger. No pain is no *carte blanche* for the maltreatment of fish (*remember the butterfly on page 64*). A fish is a living organism, an individual creature and part of the nature which sustains us. Fishing methods and handling are hugely important but so are the really hot topics like overfishing, habitat protection and restoration, sustainability and biodiversity.

Assuming fish do not feel pain means you are actively drawn into and involved with the entire process of life and not just a single facet of it (pain). What does that mean in practical terms? Imagine a once beautiful lake teeming with life now polluted, hypertrophic and close to collapse. Indigenous fish struggle to survive while an invasive species such as the grass carp thrives.

The following two scenarios highlight some of the facets involved:

1. Animal rights: assuming that fish feel pain

The ecosystem 'lake' has no moral value. The fish in it do, and their plight must be considered and their interests protected. But which species? The invasive grass carp or the indigenous fish? The majority of the fish are grass carp, so under the utilitarian calculus the grass carp have it (greatest happiness for the greatest number, best consequences for all) therefore nothing needs to be done.

That would mean, however, the indigenous fish would perish. The way out could be to allocate the indigenous fish three or five times the happiness value or assign them a different pain coefficient so as to tilt the scales in their favour. But that would mean suffering for the grass carp and the end of them. You then could say: that's not too bad: after all, death is the end of suffering. So the grass carp are, in the long run (and you're dead for a long time), better off and the lake will become useful to humans again (e.g. recreation, fishing, drinking water).

That would be a perfect, happy ending. But it doesn't happen that way. It could be argued preference for the indigenous fish would benefit humans. That is not acceptable because it's speciesist: the grass carp must not die for the benefit of humans. What's more: a healthy biodiversity would mean more life in general and fish in particular but if the latter would be fished for, that would increase pain and suffering. Intolerable. More life and fish would also attract predators, specifically, for example, cormorants, herons and crested grebes. Again this would mean an incredible rise in fish pain because there is a considerable hit and miss rate involved in bird predation of fish. Any lake restoration would be through human agency, and so it would be humans causing the fish pain and suffering. This is not acceptable. Thus with the FFP/animal rights logic, the grass carp stay and the lake collapses. You can twist and turn this story whichever way you like, the fish pain/animal rights doctrine will guide you to Absurdistan.

Lose: *the lake, the fish, the people.*
Win: *the doctrine of FFP and animal rights.*

2. Common sense, leaving aside the FFP issue

The ecosystem 'lake' has a moral existence and so have the fish and all life in it. The point here is not whether our moral obligation to life is direct or indirect (that's another debate) but that the lake ecosystem is not a moral void. That would be absurd. Thriving life and a healthy resilient fish population are good in themselves if only for aesthetic reasons! Your angling club now joins forces with other supporters of beauty and invests money and a workforce into lake restoration. The revived grass carp-free lake not only brings back beauty but also all the social and environmental benefits that go together with a healthy ecosystem. Remember the equation: water quality and habitat = fish welfare. Science and a pro-nature and human-friendly attitude are required to act on the basis of knowledge for the good of fish and humans.

Win, win, win: *the lake, the fish, the people.*
Lose: *the doctrine of FFP and animal rights.*

Out of an abundance of caution

Text books on ethics are full of fabricated moral dilemmas which you will never face and students have to get their heads around extraterrestrials who they will most likely never encounter. There is, however, no way past extraterrestrials in animal rights philosophy and activism. The argument in a nutshell is that we have to imagine that intergalactic colonisers would treat us earthlings like we treat animals. We would then stop all animal use.

PETA staged the argument brilliantly by showing extraterrestrials butchering a human and only the 'emotionally immature' are not convinced by it. Emotionally immature? This is a very popular concept used by animal rights activists. It tries to corner sceptics by suggesting that they as people are inadequate. That is if you go fishing or hunting. On closer inspection this boils down to the Soviet stratagem whereby whoever doesn't fit into the system is referred to a psychiatrist and then incarcerated in an institution. The rhetoric of *1984* is well in place.

Back to the extraterrestrials. Who else other than Peter Singer would you expect moralising at the forefront, preaching the utilitarian doctrine:

> Assuming that the extraterrestrial being is sentient, in the sense capable of experiencing pain and pleasure, and has other desires and interests that it may take us some time to ascertain, the fundamental ethical principle we should apply is equal consideration of similar interests...

But what would happen if the extraterrestrials discovered they liked nothing more than fishing and that fishing is good for them and us? Isn't it wonderful to speculate into a brainless space: you just can't go wrong.

Back to earth. For most of the time and for most people, everyday moral life is fairly straightforward: people know what is right or wrong. Unless, of course, they are utilitarian. Imagine somebody drops their wallet, and doesn't notice. What do you do? Easy. You pick it up, call for the attention of its owner and hand it back. You don't have to think about it, you just do it. You know it's the right thing to do. You act with conscience. Conscience means 'with knowledge' and I use the word in that narrow sense.

It's not knowledge but consequences that motivate the utilitarian and because of that they are heading straight for trouble. When a utilitarian picks up the wallet (assuming they do), before calling out to its owner they will have to work out what the best consequences for everybody involved are. Out of the corner of their eye they see a beggar, and in the wallet there is a lot of money, so much in fact that a lot of good could be done with it. Furthermore, the owner looks like a rich person. The loss wouldn't hurt the owner. And so on.

The salient point is that even in the most simple situations, for utilitarians there is no plain right or wrong. It just depends on the 'best consequences' for all involved – or what *you* take them to be. Anything goes, really – except, of course, going fishing.

Like everything, morals evolve but you'll find in most if not all human cultures the same values and guidelines e.g. murder is wrong, stealing and deceit are wrong, abuse is wrong and so on. The reason for this is we are all human – everybody's blood is red. There is no mystery about that. Recent research has confirmed this and in a wider sense established that seven rules are the basis of morals all around the world.

The University of Oxford website reports:

> The rules: help your family, help your group, return favours, be brave, defer to superiors, divide resources fairly, and respect others' property, were found in a survey of 60 cultures from all around the world [...] Dr Oliver Scott Curry, lead author and senior researcher at the Institute for Cognitive and Evolutionary Anthropology, said: 'The debate between moral universalists and moral relativists has raged for centuries, but now we have some answers. People everywhere face a similar set of social problems, and use a similar set of moral rules to solve them. As predicted, these seven moral rules appear to be universal across cultures. Everyone everywhere shares a common moral code. All agree that cooperating, promoting the common good, is the right thing to do.'

Common sense tells us fishing contributes to the common good because it promotes and protects healthy aquatic environments.

Recreational fishing used to be an unconditional good but by force of the FFP claim, the climate has significantly changed. With the FFP claim in hand, the animal rights movement tries to nudge people and politics towards accepting that recreational fishing is a bad thing because fish feel pain, thereby stigmatising the angler (cruel, emotionally immature etc.) and discouraging participation.

Deliberating on fishy matters is no easy task. If the 'consensus' isn't sufficient then, as already mentioned, the 'benefit of doubt' and the 'precautionary principle' are brought into play. These concepts seem moderating and reasonable – there is a feelgood factor about them – but appearances can be deceptive. The way the FFP movement invoke 'consensus' and 'benefit of doubt' is in reality a form of authoritarianism: you may or may not do this or that (i.e. go fishing).

Come to think of it, I'm all for the precautionary principle myself – and out of an abundance of caution, I say: let's distrust the FFP science and its utilitarian animal rights application. And while we're at it, let's give the angler the benefit of the doubt. If in doubt, regardless of any claims to the contrary, let's take it that fish do not feel pain.

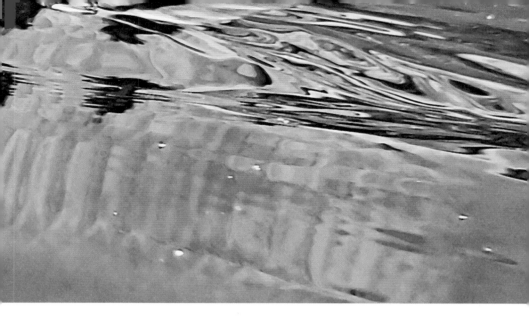

Chapter 4
Taking Stock After 20 Years

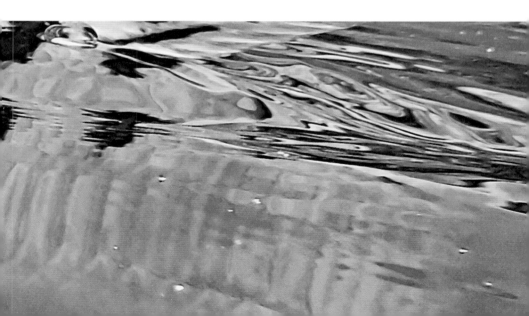

We leave the murky waters of the precautionary principle and refocus on the pain science again. The science angle is after all the key to all else in fishy matters. Intricate as the subject undoubtedly is, the key positions of the 'believers' and 'sceptics' in the major areas are clearly demarcated.

C fibre nociceptors	Believers	Sceptics
All fish:	?	No
Some fish:	Yes	Yes (very low %)
Saline fish:	Yes	Yes (very low %)

Recap: C fibre nociceptors relate the really intense pain types like a serious toothache.

Comment: Not all fish have C fibre nociceptors (e.g. sharks and rays). The cell population involved in the trigeminal nerve contains:

80% C fibre nociceptors = Full normal human pain perception.

24% C fibre nociceptors = Congenital Insensitivity to Pain (CIP). Human beings afflicted by this condition are unable to perceive pain.

4% C fibre nociceptors = Presumed fish pain.

The C fibre nociceptors are not the full story but they are central to the FFP plot. The 4% leave little doubt that a fish on the hook is not the same as a human being on the hook – not by any stretch of the imagination. Two separate research groups (Eckroth et al. 2014 and Hlina et al. 2021) confirm this point.

There are some species with and some without nociceptors. There are fish such as mantas or sharks that have no nociceptors – or they haven't been discovered yet. There are over 30,000 species of fish and only a fraction of them have been studied.

Neocortex	Believers	Sceptics
All fish:	No	No
Some fish:	No	No
Saline fish:	No	No

Recap: The neocortex is the 'pain centre' in human beings. Fish lack a neocortex.

Comment: The believers argue that other parts of the fish brain create the pain sensation and therefore no neocortex is required. Sceptics hold that those other brain parts lack the capacity to do so and are not 'designed' for that purpose. There is no need for a pain super computer like a neocortex, because of the relatively little nociceptive information that needs to be processed (C fibres).

Rocking motion and anomalous behaviour	Believers	Sceptics
All fish:	?	?
Some fish:	Yes	?
Saline fish:	No	No

Recap: The rocking motion refers to the experiment in 2003 by Lynne Sneddon in which, after a massive injection of bee venom or acetic acid, rainbow trout showed rocking motion over a gravelled tank bottom. That rocking motion is said to be 'reminiscent' of mammals (primates) in poor welfare. The fish injected with saline solution didn't show the rocking motion or anomalous behaviour.

Comment: The believers see the rocking motion as an essential part of the conclusive evidence for pain as in mammals. Neutral researchers who repeated the experiment without anaesthetic (Newby and Stevens 2008) did not observe rocking motion. The believers say they didn't properly set up the parameters of the experiment (i.e. they did not use anaesthetic or put gravel in the tank).

Avoidance behaviour	Believers	Sceptics
All fish:	?	?
Some fish:	Yes	Yes
Saline fish:	–	–

Recap: After severe, unfriendly treatment (electroshocks, temperature, alcohol, poison) fish avoid the areas where the negative experience occurred.

Comment: The believers infer from avoidance behaviour the experience of pain. Sceptics explain avoidance by innate non-pain-related abilities of fish such as learning (*see below*) taste and smell. No 'pain brain' is required for avoidance behaviour. Tardigrades and other 'brainless' animals innately know what's good or bad for them and show avoidance behaviour. Furthermore if fish pain is a 'kind of pain' but not human pain then avoidance learning is merely an external observation. We don't know what's happening inside.

Learning	Believers	Sceptics
All fish:	?	?
Some fish:	Yes	Yes
Saline fish:	Yes	Yes

Recap: Fish are able to learn – like the tank driving goldfish (*see page 47*).

Comment: If they can drive tanks, they can feel pain – they learn like us, they are like us, ergo they feel pain like us with a completely different neurology and brain from ours. That fish brain is indeed capable of amazing achievements but being able to feel pain is not a necessary condition for learning nor is conscious self-awareness. Fish like other organisms can achieve remarkable feats completely unaware of themselves, not unlike humans sleep walking or sleep driving a car.

Social behaviour and cooperation with other species	Believers	Sceptics
All fish:	?	?
Some fish:	Yes	Yes
Saline fish:	Yes	Yes

Recap: All fish show social behaviour in a way. Interaction between species is not uncommon, such as the cleaner fish and shark, or the moray eel and grouper.

Comment: The believers link the social behaviour and cooperation to a capacity for feeling pain. The sceptics say that you don't have to feel pain in order to behave socially and be communicative. Amoebas have a memory and microbes cooperate. Clever as they may be, their capacity to feel pain is not proven by observations of social behaviour and cooperation.

Pain	Believers	Sceptics
All fish:	?	No
Some fish:	Yes	Highly unlikely
Saline fish:	No	No

Recap: The saline fish in Lynn Sneddon's 2003 study did not show any rocking motion or anomalous behaviour (*see page 117*) presumed as indicative of pain.

Comment: Believers qualify their fish pain claim by saying that it is unlike human pain but still 'a kind of pain'. The sceptics highlight the flaws of the biological evidence and the conceptual difficulty of 'a different kind of pain'. Can we meaningfully talk about 'a kind of pain'?

Pain scale 1–10 and pain characterisation		
All fish:	?	?
Some fish:	?	?
Saline fish:	?	?

Recap: Humans can report the location and intensity of their pain – fish can't.

Comment: This angle is not an issue, not least because anybody would have a hard time assessing what fish pain is on the intensity side and characterising it. Nevertheless, it is a fascinating conundrum and would be a formidable challenge: you would be trying to measure something that you had no idea about. All you could assume is that fish pain is 'a kind of pain', but that is nothing to go by. Fish can't talk. If you then tried to steer away from 'a kind of pain' to, for example, 'raw experience' you would run into the exact same difficulty as with 'a kind of pain'. We just can't know because we are not fish.

All statements, presumptions and theories about fish pain are at this moment in time irreducibly human. We can try to imagine what it is like to be a fish but the more you imagine, the less fish you are.

After twenty years of FFP research the leading FFP scientists (Lynne U. Sneddon and Jonathan A.C. Roques) don't seem to be much wiser. In a recent paper they published in January 2023, they stated:

> Pain assessment cannot be generalized because it is expressed differently between individual animals. This means there is no gold standard indicator to measure pain in animals or indeed fish. Responses to painful treatment will differ between species and between individuals, as well as be specific to each type of pain.

It can neither be generalised nor specified. There is no way of telling whether a fish has toothache or a headache nor if, how and where it feels pain when it is tortured by the FFP scientists. If pain cannot be generalised or specified, probably the only meaningful way open is to research pain for each individual fish. One at a time. In that way the FFP scientists would learn something about an X number of individual fishes from different species, but nothing that applies to all fish of a species or all fishes of all species.

Affirmative fish pain science can't distinguish between mild discomfort and excruciating pain nor can it identify specific pain and its characterisation. But surely the character, intensity, duration and location of the presumed fish pain would matter because, say in aquaculture, this could help to determine welfare measures. But then again, why not turn to the pragmatic approach to fish welfare which focuses on the actual well-being of fish. Perhaps there is some Allen Carr in this situation: 'There are people who can make love standing on a hammock, but it is not the easiest way.'

There have been many studies since Lynne Sneddon's seminal 2003 article. However, nothing conclusive has since turned up. As I write, yet another article has just been published: zebrafish were subjected to the usual torture and on top of 'painlike' behaviour (not pain behaviour, but painlike behaviour!) the research team also detected a change in colour in a section of the stripes. That is very handy because with those neat stripes and the changes in it, the zebrafish can tell us *ouch*. The authors suggest using the stripes as a 'pain index'. Colour change as a result of not feeling too good is well established in popular wisdom: it's not for nothing that if you look unwell you feel 'green about the gills'.

For anglers and scientists alike the 'saline fish' is worth thinking about and keeping in mind. Anglers don't soak their hooks and bait in bee venom or acetic acid. Yet on the basis of the research by Lynne Sneddon et al. they are accused of doing exactly that. Even more remarkable: anti-fishing recommendations by ethics committees are based on this Lynne Sneddon study which actually lets anglers off the hook.

Summary:

Topic	All fish		Some fish		'Saline fish'	
	Believers	Sceptics	Believers	Sceptics	Believers	Sceptics
C fibre nociceptors	?	No*	Yes	Yes (but low %)	Yes	Yes (but low %)
Neocortex	No	No	No	No	No	No
Rocking motion / anomalous behaviour	?	?	Yes	?	No	No
Avoidance behaviour	?	?	Yes	Yes	–	–
Learning	?	?	Yes	Yes	Yes	Yes
Social behaviour/ co-operation	?	?	Yes	Yes	Yes	Yes
Pain	?	No*	Yes	Highly unlikely	No	No
Pain scale 1–10	?	?	?	?	?	?

* Elasmobranchs (sharks and rays) are fish, but they do not even have nociceptors, and therefore cannot feel pain.

The Manifesto: the emperor's new clothes

'Manifesto' rings a bell: 'The proletarians have nothing to lose but their chains. They have a world to win. Working men of all countries unite!' Whatever happened to the working class? Did it wither away? No, the intellectual smart set has brushed the proletarians under the carpet and moved on. The high-minded savants of animal rights don't ponder on real questions like wages, pensions, housing, health insurance, fishing licence fees and the like. There is no kudos to be gained from unglamorous necessities whereas to champion animal rights practically guarantees you a spot in the limelight.

The end of all animal exploitation

Think about it: 'the end of all animal exploitation'. Goodbye to all the companion pets, the fishing, the pony riding, the guide dogs... all the human interaction with animals. Even a seemingly innocent activity like bird watching from a hide is just not on. The feathered beauties are 'fully-fledged members of our society' so they surely have a right to privacy but probably worse than violating their privacy is that you enjoy watching birds. You abuse the poor animals for your own entertainment. Likewise enjoying birdsong is a no-go zone because your spirits are being uplifted by undiluted male aggression and territorial behaviour.

You need to undergo some serious brainwashing if you want to participate in the 'end of all animal exploitation'. 'The end of all animal exploitation'. That's goodbye to lactose in pharmaceutical products, and leech therapy in alternative medicine? It has to go, just like alternative foods obtained from insects or mealworms because they are sentient too. We should allow rat infestations everywhere (fundamental rights for rats), in hospitals, in every home. And if you follow this Manifesto to the letter you end up starving yourself to death. Not that the authors and signatories would apply these rules to themselves: far from it. After all they are currently very happy to use animal products (as used in batteries, screens, food, books) or exploit

Manifesto for the Liberation of the Animals

There is no social change without a strong social movement. We, from the **animal rights movement,** stand in the tradition of the great liberation movements and demand **fundamental rights for all sentient beings:** The right to life, to freedom and to integrity.

This implies **the end of all animal exploitation!** Be it for research, for food, for entertainment or for other human purposes. A little more straw or a little more space will not do away with our injustice done to the animals. We need visions on how to live with the animals: They must no longer be objects and victims of our violence, but must be recognised as **fully-fledged members of our society.** The road leading to this vision is long and full of obstacles. However, we are ready for it: determined, unfaltering, nonviolent.

Our focus is on the animals. However, it is not solely on them. For discriminating against one group often helps to support the oppression of another. Hence our demand for **justice and solidarity** is a broad one, and it encompasses all social struggles against oppression and discrimination. Species, gender, ethnicity or any other arbitrary criterion must **never** be a reason to enslave or exploit a being.

We urge you: Be creative and develop visions of an equitable coexistence! **Form alliances and be outraged!** Join our movement and fight **for the liberation** of all sentient beings!

animals (as pets) whenever and wherever it suits them. The usual story: preaching water and drinking wine.

The authors of the Manifesto are, as you would suspect, highly educated, First World brains. The Manifesto website asks you to sign in support. I would fully understand if you said 'I'll never sign such baloney'. Nevertheless, at the moment there are about 1,800 signatories who have publicly signed their support. Again the same picture: well-known and acclaimed university professors like Gary Steiner, Marc Bekoff and others.

And then there is Philip Wollen. Philip Wollen is an Australian merchant banker turned animal rights activist rubbing shoulders with Peter Singer, David Attenborough and the likes. In no time at all the Manifesto punches way above its weight. There is another reason for mentioning Philip Wollen: there are quite a few rich, prominent animal lovers like him around, financing animal rights causes and research. Patronage for art, science and all kinds of issues is as old as culture and there is nothing wrong with that. It's all part of the competition of ideas and that is a good thing. It just seems that the scales are at the moment a bit heavily tilted to the *1984* side.

The manifesteers are 'determined, unfaltering' and 'non-violent' yet at the same time they 'fight' for animal liberation. What is a non-violent fight? Whatever exactly they mean by 'fight', fight they do: sometimes with real violence on the river bank, with verbal violence at conferences and with abusive and threatening hate mail. Not the instigators though: the dirty work is done by the boots on the ground such as ALF (Animal Liberation Front). It's gone quiet recently but it's still around.

And even if they don't throw stones at you or in the river when you're fishing, you are at the receiving end of clever guilt-tripping. They want to make you feel bad about what you eat, what you think and how you live. It's constant, joyless pressure. All over the First World it is the same kind of people who try to take as much joy and fun out of your life as they possibly can. Incessantly repeating the message at all levels in all possible channels, with lots of Hollywood stars giving it credence too, they will eventually deliver the desired results i.e. animal rights laws.

This Manifesto is a great testimony to the authors' and signatories' PR and business acumen. They judged correctly that they would get away with this nonsense and they did. It takes guts to step out in the emperor's new clothes. Full marks for that.

2 + 2 = 5

The killjoys don't shy away from dismantling anything which is in the way of a just and better world. Nobody and nothing is safe. Just imagine if the decimal system had been invented by some badass colonialist who hunted and fished. Would we have to decolonise maths, and tear it down like the statues of the colonial oppressors and erase its use and history? What would the new decolonised maths look like? Would the new maths make 2 + 2 = 4 any less true than it is?

The decimal system originated in ancient India and travelled from there with the help of the Arabs to Europe which is why our current decimal system is called the Hindu-Arabic system. It travelled with trade and war. The Moors conquered and ruled the Iberian peninsula between AD 711 and 1492. You could make a really good case for saying that the Moors initiated modern European development in science and philosophy. They brought with them advanced mathematics; and the paramount influence of Aristotle on European philosophy would have never happened without the mediation of the great Islamic scholar Ibn Rushd (aka Averroes, AD 1126–1198).

So far so good: I am on safe, politically correct, inter-faith, gender neutral ground here. This probably changes now with the question: Does anybody want to tear down the Alhambra in Granada? After all, it's a monument documenting and celebrating the superiority of one culture over another, namely the Moors over the then-Iberian locals. Or look at the role and treatment of women in ancient India and the structure of Islamic society at the time (women, slave trade). Does it really make a difference that it happened so long ago? And, again, does it mean we have to decolonise the decimal system?

All science is influenced by the social, political, economic and intellectual context of the day. For decades the Western ethical elites have been hellbent on deviating as far as possible from common sense. Truth – be it in mathematics or in concepts such as beauty – currently has a bumpy ride. In order for 2 + 2 = 4 to be remotely true, it has to run the gauntlet of:

- decolonisation
- gender
- sexual orientation
- race
- positionality
- alternative ways of knowing
- personal experience
- citation justice
- social justice
- democratic legitimisation (majority decision)

And in the end 2 + 2 = 4 has to be signed off by the Diversity and Inclusion Officer and a representative of the Ministry of Truth. Should the politically and ethically correct powers that be for some reason decide that 4 is not the desired result, then it's going to be 5. To paraphrase George Orwell:

> In the end the politically correct will announce that two and two made five, and you would have to believe it... the heresy of heresies is common sense.

1984 is not far off. In fact we might already be there.

Either way, Orwell is as relevant and important as ever:

Freedom is the freedom to say that two plus two make four. If that is granted, all else follows.

Chapter 5

'Wherever The Trout Are, It's Beautiful.'

This is a most remarkable and inspiring observation by the Czech statesman, philosopher and master angler Tomáš Masaryk (1850– 1937). Imagine now the perfect pastoral idyll: a gently murmuring stream meandering its way through the leas. Bees buzzing, birds singing, trout leaping and so on… this Arcadian scene informs us that the water is of great quality.

The trout there mean there is sufficient structure and cover for them, despite the herons they attract. The trout also tell us that there is a rich insect life, and the birds jubilantly confirm this. The birds themselves are of interest to small predators, while the bees are enjoying the nectar of the wild flowers. The presence of trout indicates also that riparian agriculture is of a gentle kind, anglers are not

overfishing it, and that the high-tech sewage treatment plant further upstream works fine and is neutralising the endocrine disrupters *(see page 69)*.

The trout affirm not just the healthy state of the water but of the entire habitat and local environment. They reveal the beauty in everything else around them. It needn't be a pastoral idyll: it can be a village stream or pond. It needn't be trout either: in different climates in different water systems, it is different species that make you see beauty.

The point to note is that natural, pollution-free waterscapes will always be aesthetically pleasing, be they streams, rivers, ponds, lakes, tidal zones or the oceans. Water initiates and sustains life everywhere, and life in turn creates beauty. Protecting water is protecting beauty, restoring water is restoring beauty, conserving water is conserving beauty.

'Beauty' is a giant topic which I shrink to a few lines. The 'beauty' Masaryk had in mind was the 'beauty of nature'. The question is: what is beauty and what is it in nature? It is often said that beauty is in the eye of the beholder – that is to some degree certainly the case. However, that doesn't make it entirely subjective, nor does it exclude the possibility of it being objective and really existing. Beauty, like truth, could be independent of any human agency. If $2 + 2 = 4$ is true, it is true independent of human awareness of the fact. For thousands of years people didn't know there were dinosaurs yet they objectively existed in their time – people knowing or not doesn't come into it. Of course contemporary sociologists, evolutionary biologists and philosophers doubt, fragment, attack or deny beauty with great gusto. On the other hand there are physicists, mathematicians and biologists who can see beauty in their fields and talk of it as a given. The long and short of it: beauty is not esoteric knowledge but an experience many, if not all humans, share.

Such may be the case but it doesn't explain or at least give you a vague idea of what natural beauty could be. Apart from its particular occurrences, beauty is always the same thing. You can look at completely different environments or scenes in nature, and beauty is always the same phenomenon. Be it an insect, a flower, a tree or a trout, you are looking always at an interplay of symmetries, harmonious proportions, and self-similarity, and more often than not the Fibonacci numbers are involved. The Indian poet and mathematician Pingala (second century BC?) and the Italian Fibonacci (AD 1170–1250) discovered that

$1 + 2 = 3$
$2 + 3 = 5$
$3 + 5 = 8$
$5 + 8 = 13$
$8 + 13 = 21$
$21 + 13 = 34$
...
$2584 + 4181 = 6765$
...

and so on. Now divide 3:2, 8:5, 13:8, 21:13… 6765:4181 and you get closer and closer to the phi of the golden ratio, which is 1.618.

These relations are graphically summarised in the Fibonacci spiral:

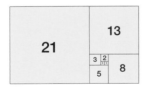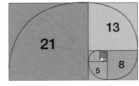

Be it DNA or galaxies, pine cones, sunflowers, sea horses, tree branches, leaves, hurricanes, meandering streams or fish, wherever you look in nature you're bound to stumble over Fibonacci numbers and the Fibonacci spiral. A particularly fascinating observation is that the leaves of many plants grow in line with the Fibonacci spiral because thus the available space for each leaf is maximised for photosynthesis. And by the same token, when it rains, the water spirals down – with minimum loss – to exactly the right area to water the plant.

What we recognise as 'beautiful' say in a rose or any other flower is something both out there and in us. It's not just a random impression. When we see, feel or talk of beauty, we talk of something real, something which exists out there. As mentioned above, some of the objective properties of beauty are symmetry, harmonious proportions, self-similarity and Fibonacci.

The pastoral idyll is packed with mathematics, which is why I said above that beauty, like truth, can be something completely independent of human perception. However, there are more than numbers in Arcadia: there are scents, colours and sounds. And there is animal life of all kinds. When we look at 'beauty' we look at literally quadrillions of events happening at the same time. Just think of all the chemical processes taking place in plants as we enjoy the colours and smells of the meadow flowers and herbs. Think of photosynthesis. Think of all the flies hatching, all the little critters hurrying around, all the bees, birds, the mice and hedgehogs, and not least the trout, those red speckled beauties in the water chasing around for food. There is biodiversity for you: Arcadia is not monoculture but diversity – the trout stream is teeming with life with all its facets buzzing with activity. Beauty is a process, beauty is the process of life.

If you experience beauty, you're perceiving life unfolding in all its diversity. Although it's practically an infinite number of events you're seeing, you're seeing them as one: unity in diversity. You don't need to know or think that such is the case; it's your aesthetic autopilot which does that for you. If for example you look at just one meadow flower, you're perceiving also all the pulsating life around you. The flower you're looking at is not a single item. It is part of a package and although you might be just looking at the one flower you are at the same time aware of all the processes around you.

Far away from the Masarykian world of trout, on tennis courts something similar is happening as far as perception is concerned. The really great players such as Roger Federer have their eyes always on the ball. At the same time they are aware of the position of their adversary, the light, the wind and the audience. For his volleys, slices and lobs, Roger doesn't need to think 'ball there, me here, racket in right hand, Rafa to the left, now play dropshot' – his autopilot does all that for him. Rafa and Roger hit the ball not blindly but with all their senses open, yet they are focused on the ball only. If you look at the flower, you look at everything around you simultaneously. Anglers experience this too. Sometimes you just know that the next cast will catch. As the line flies out you know this is it, although you have no conscious information why that should be. It's not extrasensory perception but probably ultrasensory perception. Paraphrasing Shakespeare I conclude that there is more to my brain than can be dreamt of in my philosophy. Hopefully. One more thing before we move on: if we humans are not supernatural or extraterrestrial but part of nature, then we are also part of beauty, part of the process of life. Experiencing beauty in nature means being alive, being part of life and enjoying life.

Truth is old hat?

The above sketched view of beauty borrows heavily from the classical model which aligns beauty with truth and goodness. These days such a take won't make many friends, but then who needs friends who cultivate ugliness? The very idea of 'truth' is suspect, and beauty and goodness provoke ridicule – or worse – because they could be part and parcel of ideology that could be inimical to the current all-inclusive anti-this-that-and-the-other dogmas. Breaking a lance for truth is Albert Einstein (1879–1955) in a fascinating conversation with the Indian poet and philosopher Rabindranath Tagore (1861–1941):

TAGORE: ...Science is concerned with that which is not confined to individuals; it is the impersonal human world of truths. Religion realizes these truths and links them up with our deeper needs; our individual consciousness of truth gains universal significance.

Religion applies values to truth, and we know truth as good through our own harmony with it.

EINSTEIN: Truth, then, or Beauty, is not independent of Man?

TAGORE: No.

EINSTEIN: If there would be no human beings any more, the Apollo Belvedere would no longer be beautiful?

TAGORE: No!

EINSTEIN: I agree with regard to this conception of Beauty but not with regard to Truth.

TAGORE: Why not? Truth is realized through man.

EINSTEIN: I cannot prove that my conception is right, but that is my religion.

TAGORE: Beauty is in the ideal of perfect harmony which is in the Universal Being. Truth the perfect comprehension of the Universal mind. We individuals approach it through our own mistakes and blunders, through our accumulated experiences – through our illumined consciousness – how otherwise can we know Truth?

EINSTEIN: I cannot prove that scientific truth must be conceived as a truth that is valid independent of humanity but I believe it firmly. I believe for instance that the Pythagorean theorem in geometry states something that is approximately true independent of the existence of man.

If there is truth independent of human beings, why shouldn't the same apply to the beauty of nature because nature is full of mathematical properties and so is the Apollo sculpture. The English poet John Keats (1795–1821) would probably endorse that thought judging by the two concluding lines of his famous 'Ode on a Grecian Urn'.

Beauty is truth, truth beauty,—that is all
Ye know on earth, and all ye need to know.

Is beauty old hat? It sure is if we are to believe its modern detractors in art and philosophy. On the other hand, it sure is as old as humanity and certainly worth casting a line for with Einstein or Tagore. Either way, with beauty you're in for an inspiring experience and you can only win!

Sister act

Beauty was always there. We don't know what our distant ancestors made of it but we know that the beauty in nature emerged as a poetical and practical theme in the nineteenth and early twentieth centuries. Not least because ugliness in the big cities became unbearable. Big city – big stink. This was especially true of London, though not exclusively so. The Thames was 'thick with human sewage'. The condition of the working classes and the poor was just as ugly as the stink of the towns themselves.

Miranda Hill (1836–1910) founded the Society for the Diffusion of Beauty (1875) and was joined by her sister Octavia Hill (1838–1912). The aim of the society was to 'bring beauty home to the poor'. Part of that beauty was space and natural beauty. The society had apparently some initial success but then stalled. However, Octavia persevered and eventually became one of the founding members of the National Trust and her legacy thus lives to this day. The National Trust website quotes her with: 'We all want quiet. We all want beauty… we all need space.' The space and the natural beauty delivered respite, fresh air and distraction. Apparently Octavia coined the expression 'green belt' which tells a story by itself. What an extraordinary woman. The National Trust website sums up Octavia's pioneering achievement as:

> Octavia became convinced of the need for open spaces for the urban masses. She joined a campaign to save Swiss Cottage Fields from development and although it eventually failed, it was through it that she met Robert Hunter, solicitor for the Commons Preservation Society.

They successfully campaigned together to resist development on Parliament Hill Fields, Vauxhall Park and Hilly Fields in London. Ultimately, along with Hardwicke Rawnsley, they went on to found the National Trust for Places of Historic Interest or Natural Beauty in 1895.

For the next 17 years until her death in 1912 Octavia continued to fight for the preservation of the countryside. She helped the National Trust to buy and protect its first land and houses and campaigned for the preservation of footpaths to ensure everyone had right of access to the land.

An interesting aside is another remarkable woman with a completely different background and biography. In 1905 Marguerite Burnat Provins (1872–1952) an artist and author founded the League for Beauty (*Ligue pour la Beauté*) in order to protect Swiss nature from 'vandalism'. Part of the fight against vandalism was the protection of the indigenous flora and fauna. The League for Beauty was a precursor of modern Swiss nature conservation. Wherever you peek a bit into the history of nature protection and conservation, the aesthetic angle isn't far off. Beauty is a powerful driver for conservation and these days that also means business. Fundraising for environment NGOs would be a lost cause without beauty. Far from being an elusive entity, beauty is a fact in the real world.

Is beauty legal?

In Ecuador, Bolivia, India, Bangladesh and New Zealand rivers have rights. Not all rivers but some of them. These rights were granted for different reasons, but two of them are in common: pollution (ugliness) and the insight that a river is a living entity. The New Zealand legislation states: 'Te Awa Tupua [Whanganui River claims Settlement] is an indivisible and living whole, comprising the Whanganui River from the mountains to the sea, incorporating all its physical and metaphysical elements.'

On the basis of rivers being living entities, they are granted 'legal personhood' which is not the same as the personhood discussed in animal rights. In animals rights 'personhood' means a natural person like you and me and 'rights' means moral and legal rights. A legal person is different. A legal person is a legal entity, for example, a corporation, which can also have 'rights' but as that entity. According to the law the Whanganui River is one such legal entity. The river, like the corporation, can't speak for itself nor appear in court. Its interests and rights are looked after by appointed guardians who, like the executives of a corporation, act on its behalf. As you would guess, this view is not entirely undisputed.

'Pollution' apart from the obvious sewage and industrial waste, might include dams or runoff from industrial or agricultural areas or infrastructure. Pollution doesn't need any further explanation. The 'living entity' on the other hand is worth having a look at. I don't think anybody would seriously contest that rivers are living entities. Ideally they are fully alive as healthy ecosystems and if not they are hopefully restored and revitalised. Go a step further and look at the whole picture: the Gaia hypothesis holds that the entire planet is a single, self-regulating system – not a thing but an organism.

Back to beauty in nature: healthy ecosystems produce natural beauty – they can't help it. It's literally in their nature. By protecting beauty, you protect ecosystems and by doing that, biodiversity. Natural beauty doesn't necessarily require wildernesses such as the Amazon Basin. Wildernesses in the sense that they are untouched by humans don't exist any more. Besides, as mentioned above, humans are part of nature. Natural beauty and rich biodiversity are also found in cultural landscapes. Traditional extensive Swiss alpine farming over time has created a unique environment that is endangered when alpine pasturing is given up because it's not profitable any more (price of milk). In areas where summer grazing has stopped, the consequences are obvious: shrubs and bushes take over and biodiversity declines dramatically – mainly affecting plants, insects and birds.

Now remember our ethically motivated vegan and vegetarian friends. They do nothing to preserve this alpine biodiversity by refusing to consume any animal products. They can't even enjoy wonderful Swiss chocolate because of the milk in it. You on the other hand have a duty...

Natural beauty doesn't need vast expanses. A flower pot on a balcony on the twentieth floor, a little lawn with a tree – wherever there is a little bit of living nature there is beauty to see and enjoy. Urban parks are spaces where people can connect to beauty. The contribution of urban parks to mental and physical health is well documented and so are their beneficial effects regarding noise and micro-pollution and the cooling effect during hot weather. Beauty, be it on a large or a small scale, is also utility. In the bigger picture, beauty provides the so-called ecosystem services including protection (erosion, avalanches, flood control), water purification (healthy environment, healthy water), climate regulation (healthy air) and carbon storage. And as with on the balcony or in the park, healthy ecosystems and biodiversity provide pleasure and recreation.

Beauty is also business: I already mentioned fundraising, but think of tourism and, for example, of engineering (sewage technology can help to maintain beauty). Instead of legislating for rivers you could also legislate for beauty which in some jurisdictions (Germany, Switzerland, England, Wales and Scotland) is already being done. Scotland defines its natural heritage as: 'the flora and fauna of Scotland, its geological and physiographic features, its natural beauty and amenity.'

Even the dry Scottish lawmakers couldn't ignore beauty, not least because millions of tourists visit Scotland for its sublime natural beauty. Is beauty legal? It sure is. Even if it wasn't, pollution is a crime against beauty. The long and short of beauty is:

- Beauty is the process of life
- Beauty signals health and biodiversity
- Beauty is not esoteric knowledge but an experience all humans share

- Humans are part of nature and part of its beauty
- Beauty is accessible to all
- Beauty also means utility and business

Why is this beauty business so important? Throwing off the shackles of pain and looking at beauty puts the FFP question into perspective and allows for consideration of the whole picture.

Pain or beauty?

Deciding to view the world in terms of pain or in terms of beauty illustrates what I mean by that:

1. The advocate for the pain lens is none other than Peter Singer himself:

> Fish also show signs of distress when they are taken out of the water and allowed to flap around in a net or on dry land until they die. Surely it is only because fish do not yelp or whimper in a way that we can hear that otherwise decent people can think it a pleasant way of spending an afternoon to sit by the water dangling a hook while previously caught fish die slowly beside them.

This is great billboard stuff: the flapping and slowly dying fish go straight to the heart. Having said that, if the angler wants to keep a fish, they kill it right away.

2. A well-known advocate for beauty is the American poet, philosopher and angler Henry David Thoreau (1817–1862). He wrote:

> Ah, the pickerel of Walden! When I see them lying on the ice, or in the well which the fisherman cuts in the ice, making a little hole to admit the water, I am always surprised by their rare beauty, as if they were fabulous fishes, they are so foreign to the

streets, even to the woods, foreign as Arabia to our Concord life. They possess a quite dazzling and transcendent beauty which separates them by a wide interval from the cadaverous cod and haddock whose fame is trumpeted in our streets. They are not green like the pines, nor gray like the stones, nor blue like the sky; but they have, to my eyes, if possible, yet rarer colors, like flowers and precious stones, as if they were the pearls, the animalized nuclei or crystals of Walden water. They, of course, are Walden all over and all through; are themselves small Waldens in the animal kingdom, Waldenses.

The book *Walden* is a milestone in American literature. Its title, *Walden*, refers to Walden pond in Concord, Massachusetts. On the shores of Walden pond, Thoreau lived an ascetic life for two years in a simple cabin he built himself. Thoreau's lens is not fogged up with pain.

The messages from Singer and Thoreau couldn't be more different. Now I may be accused of comparing apples with pears because Singer (politics, ethics) and Thoreau (beauty, nature) wanted to convey different messages. Yes, that is the point: the different messages. So now let's look at the message of beauty in fishing and what it means in practical terms.

The quest for beauty

Going fishing is the quest for beauty and more often than not you'll find it because beauty is irresistible if it's there. Natural beauty has a subtle way of winning you over, breaking down the defensive barriers of the daily grind.

Under the umbrella of beauty the mind wanders, you contemplate, you focus effortlessly on the essential, time seems to disappear and even thinking or preparing for fishing gets you almost there. The fishing rod is a bit like a magic wand – just holding one gets the dopamine flowing. When you're in the tackle shop, taking that beautiful rod you absolutely don't need out of the rack, feeling it and judging its balance, you're already elsewhere – you're fishing even when you're not. Can you look at a float in the tackle shop without seeing it bobbing? Even thinking about fishing literally takes your mind off all other things.

Fishing draws you deep, deep into the works of nature and beauty. Fishing naturally engages the angler with water and with water quality. Likewise anglers inevitably find themselves interested in biology, geology and geography. If you go out with a boat, all sorts of skills and knowledge are required, just as you need to be prepared for all sorts surprises when heading for remote alpine lakes or Scottish hill lochs. In an age where everybody is a kind of specialist or expert, the angler is the exception: anglers are generalists. In order to be successful they need to draw on all kinds of knowledge and resources. The angler is a true *uomo universale* (*donna universale* too, of course) – the Renaissance ideal of the individual who excels in science, arts and action. Less flatteringly: a Jack (or Jill) of all trades and a master of some (otherwise no fish caught).

Many areas of human research and knowledge play into fishing. Space science is part of fishing. In fact some equipment and outfits are so sophisticated you could probably land on the moon with them. If you compare pictures of anglers over the last two centuries, the contemporary angler (especially the fly fisher) looks a lot like an astronaut landed by the riverside.

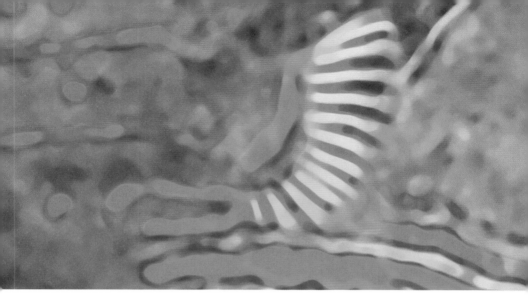

Chapter 6
How Do You Communicate With Fish?

Drop them a line. There is more to this well-worn joke than meets the eye. The trilogy of hook, line and knot merits a closer look. After all this is how it all started about 20,000 years ago. The hook was discovered, not designed! The potential of suitably shaped pieces of wood or bone to hook a fish must have been recognised. But that by itself wouldn't have been sufficient. Only with the help of maybe a flexible stick or a line woven with some plant material (e.g. flax) could fish be caught.

Nowadays hooks are high-tech objects in terms of material and quality. Their precision and smoothness and functionality make them aesthetically pleasing objects. The angling hook is not a tool like a hammer or a knife. It's not a weapon either. It's an object which on its own is useless. Only the line, the knot and the angler can make it work. The ensemble of hook, line and knot was a decisive achievement in human history: it was the beginning of the modern world and the moon landing. Take one element out of the basic fishing equation and you never land anything at all – only the faultless interplay of the basic components and a savvy fisher will catch fish.

For centuries the fishing hook has been rooted in the collective unconscious. Everybody understands the concept – it's a powerful visual. This is the reason why the hook is such a fantastically efficient marketing tool for the FFP/animal rights movement which uses it very cleverly. Unlike oestrogen, benzodiazepines and nanoplastics (to name but a few pollutants) the hook needs no explanation. It can easily be portrayed as the source of all evil that befalls fish and conveniently blindsides the bigger issues of fish welfare (pollution, habitat destruction).

Flashback

The key visual of much anti-angling propaganda is the hook. Great play is made on how painful that is to the fish (similar to a human being or a dog being hooked). The scientific reality tells another story. Remember that famous experiment by Lynne Sneddon in 2003 and those trout injected (lips!) with various substances:

- Bee venom
- Acetic acid
- Saline solution

One group of trout was not injected with anything at all (the control group). On the effects of those substances Lynne Sneddon's study reports:

> Observations following acid and venom injection showed that the fishes performed anomalous behaviours after the treatment that were not seen in the control or saline groups...

The bee venom and the acetic acid produced a technical knockout by stunning the entire nociceptive system of the trout. But look at the saline solution group: no anomalous behaviour after the syringe delivered the saline solution! The penetrating needle of the syringe is the equivalent of the hook – it is the same, except that anglers don't dip their hooks in saline solutions, venom or acetic acid. Think. You could say that this is a case of *the truth hidden in plain sight.*

Nevertheless, the evil angler hides the nasty hook in food. Although it is the case that for example perch will have a go at an empty shining hook, this isn't the rule. Fishing is about seduction and rule number one of all seduction is that the bait must be attractive to the fish and not the angler. Obvious? Well, fly fishers in particular seem to be prone to fooling themselves rather than the fish with their pretty flies.

Like the hook, the basic form of the line can't be improved. A line is a line and by itself is useless because you don't catch a fish with just the line. But the line is indispensable as the connecting medium, and in that respect nothing has changed since the first line. However, in terms of the material there are worlds between flax, raffia, horsehair, gut, braided silk and biodegradable fluorocarbon monofilament.

Knotty business

There are good knots and there are bad knots. The best knot is the one you have confidence in. The source of that confidence is your application of knotting knowledge. Not that the angler needs to delve into mathematical knot theory (but then again, why knot?!) but not knowing about the knot is not OK. You have to be serious about your knots. As Roland Pertwee observes:

> Your true born angler does not go blindly to work until he has first satisfied his conscience. There is a pride in knots of which the laity knows nothing, and if, through neglect to tie them rightly, failure and loss should result, pride may not be restored nor conscience salved by the plea of eagerness.

In the *History and Science of Knots* (1996) a knot is defined as: '…a single closed curve that meanders smoothly through Euclidian threespace without intersecting itself.' I haven't got a clue what that means but when it comes to a tangle, be it a simple wind knot or a kingsized tangle, it doesn't seem to be helpful. A simple wind knot which in reality is a casting knot (bad casting) looks like a minor annoyance but it is a nuisance for aesthetic reasons, like a blot on the landscape; and it also weakens the line. The tangle is a truly knotty fishing issue. There seem to be three laws governing tangles:

1. It's dead easy to get yourself into a tangle.
2. While you disentangle, suddenly the fish are on the take left, right and centre.
3. Once disentangled, the fishing action is over.

While there is consensus that this is broadly speaking the case, some additional observations might be conducive to better understanding.

HOW DO YOU COMMUNICATE WITH FISH?

Take the classic case of the seemingly simple tangle that you hope to unknot by giving it a gentle pull at one end. This results either in a tight knot or like magic in a new tangle of mind-boggling dimensions. In fact I suspect that getting from a tangle into a worse tangle is often easier than getting into the initial tangle. Similar to this situation is the line wound around your rod tip, which is a kind of baby tangle. With a few flicks you try to unwind it. The very moment you do that, a gust triggers off some seriously foul language.

The tangle laws seem the curse of the angler. However, there is no law saying you couldn't look at tangles from another angle. Are they really such a curse? It's perhaps just a question of attitude: don't be so negative! Why not opt for an aesthetic approach? Look at the tangle and instead of an annoyed 'Oh, bugger,' go for an enthusiastic: 'Wow, what a beauty!' And why rush into disentangling? This little beauty was blown together exclusively for you. By assessing the tangle in an aesthetically positive way you can understand its intricacies better and identify more clearly the best course of action. Not only that. If you let yourself be negatively taken in by the tangle, you get tense or upset which then clouds your powers of observation and judgement. By going for the aesthetic approach you come out of the tangle relaxed and ready for action again.

The lens of beauty: business

Beauty means business: all the tackle, travel, hospitality, boats, literature and other gear of the First World's estimated 220 million anglers (2023) is phenomenal. That is the direct economic benefit. The indirect benefits, such as improved mental health, are difficult to quantify. While the worldwide impact can be tricky to estimate, locally there are some fairly reliable figures.

The Australian Department of Agriculture, Fisheries and Forestry reports on the results of a recent national survey (2023):

A national social and economic survey of recreational fishers has shown that one in five Australian adults participate in recreational fishing every year, improving well-being and contributing 100,000 jobs and $11 billion to the Australian economy [...] Recreational fishers were found to have, on average, higher levels of well-being than non-fishers, and those who fish more often have higher well-being. Recreational fishing appeared to support positive social connections, nature connection, relaxation and can help achieve recommended levels of physical activity.

Figures vary from country to country and from year to year but the impact of recreational fishing on First World economies is substantial. The Australian survey also established that 79% of Australians have a 'positive outlook' on recreational fishing 'considering recreational fishing to be an acceptable activity.' I am not following up the obvious question of the negative 21% because we are looking at fishing now through the lens of beauty.

If anglers were to go on strike and stop their volunteer work and free services the world would surely be a poorer place. Not many countries could afford to pay for the services delivered by volunteer workers. I just pick two random examples from the internet to hint at the scope. *The Guardian* (UK) reported in 2022:

Anglers, of which there are at least 2 million in England, go down to their treasured slices of waterway whenever they can to tend them, trimming vegetation, creating wetland spawning habitats, and even painstakingly cleaning the gravel. It sounds like a pretty peaceful pursuit, but when the Guardian went to visit some Angling Trust members at their clubs around Reading, there was palpable anger in the air.

This is because water companies have been spewing waste into many of these stretches, destroying the hard work, money, and hours of time anglers put in to keeping the rivers healthy. Now, they are fighting back with determined fishers all over the country testing their stretches of river for pollution using kits supplied by the Angling Trust. Often no one else will do it...

Note: healthy canals, streams and rivers benefit everybody, not just fish and anglers.

The scope of volunteer work ranges across pretty much everything which serves beauty. In this example it is collecting data for science. Another random example picked up from the internet is from New Zealand:

A successful Waitaki River hatchery finclipping morning was recently held. On Easter Sunday 65 volunteers turned out and 15,549 fish were fin clipped in less than two hours.

Literally hundreds of thousands of volunteer activities take place all over the world wherever anglers are active. Volunteer work doesn't need to be angling related: angling clubs often team up with other local clubs to organise charity events or the village festival.

Last but not least, rod licences need to be mentioned. Under the heading of: 'Fishing makes the world a better place' an American website puts it succinctly:

The fees you pay for a fishing license go directly to fund water and land conservation programs in your state and throughout the United States. More forests and cleaner rivers, lakes, and oceans mean more opportunities for you to tap into the free medicine that fishing and surrounding yourself with nature represent. It's a loop of positive reinforcement.

That applies in principle the First World over. Incidentally, 'Fishing makes the world a better place' is an absolutely brilliant summary of looking through the lens of beauty. It sums up what recreational fishing is all about.

The lens of beauty: joie de vivre

Evidence from the world over testifies to what beauty and fishing can achieve. Apart from the obvious benefits of fresh air, exercise, honing motor skills, there are other areas where fishing creates joie de vivre. Wheelchair fishing is one such area where over the last decades remarkable progress has been made. It's based on anglers' initiatives and volunteer work and covers all kinds of fishing (stream, river, lake, sea). It is happening all over the world and again I cite just one random example. Fishtalkmag.com reports:

> In a nutshell: the *Redeemer* is a 46-foot Composite Yachts Chesapeake Bay deadrise, custom designed and built for wheelchair accessibility… It is, to our knowledge, the only working charter boat with full 100 percent wheelchair accessibility and ease of use, designed and built specifically to take wheelchair-bound and mobility-challenged anglers and their families out onto the Chesapeake Bay for a day of fishing… The Fish Redeemer team didn't come up with the idea of a wheelchair-accessible charter boat to make money, they did it to help people who wouldn't normally have the opportunity to go fishing on a boat find the ability to break through this barrier. They had always loved fishing these waters. When they found that some people were unable to enjoy the experience due to mobility issues and that no one around was offering this service, the team and its board of volunteers created Fish Redeemer.

On a more general level, an Australian study by Professor A. McManus, Dr. W. Hunt, J. Storey and J. White concluded:

> This study found that considerable health and wellbeing benefits can be gained through involvement in recreational fishing. Encouraging young children, youth, adults and families to fish offers healthful outdoor recreational activity that can be enjoyed throughout life. Benefits were evident for individuals

and groups. Recreational fishing also provides significant benefits to children and youth with behavioural and mental health issues.

The major benefits identified were: youth development; social support; good mental health outcomes; behavioural management; rehabilitation of upper body musculo-skeletal injury and reductions in stress and anxiety. Seniors can also gain significant health benefits by continuing to remain active both physically and mentally through this enjoyable, low cost outdoor pursuit. Intergenerational transfer of knowledge and skills from seniors to younger generations is another major benefit that should be exploited by recreational fishing groups.

That sums it up neatly. Nevertheless I want to emphasise one more angle. Young people who are into fishing and who have got into some kind of trouble usually have adult friends they trust more than their parents. Fishing can provide the space and opportunity for positive problem-solving. Besides, angling literally keeps the kids off the street – they can't be in two places at the same time.

My final point is that fishing makes a better world also for those with ADHD. ADDers.org.uk reports:

Research is proving that fishing can be a natural remedy for symptoms of Attention Deficit/Hyperactivity Disorder (ADHD). One study surveyed parents of 500 children with ADHD and found that simply spending time in the great outdoors while casting out a line can serve as a buffer for depression, anxiety, weak immune systems, and other psychological and physical factors of ADHD. Fishing can also teach children with ADHD to manage their restlessness, focus their attention, and stay calm during a fun activity.

There is no point in citing more examples. The evidence is overwhelming: fishing is good for you – if ever you doubted it.

The lens of beauty: aesthetic experience

There is no such thing as a free lunch. Fish learn this too when they go for your bait. Likewise the fishing experience doesn't come completely free: it is what you make of it i.e. what you 'invest' in it. You can be a happy angler with elementary knowledge about entomology and you can be equally happy as an authority on the subject. There is no one size fits all – you can, for example, be the most complete of anglers without ever reading a book about fishing.

You can make volunteer conservation work part of your fishing. This will no doubt expand your horizon and add more pleasure to the experience. Working in the pursuit of something beautiful rubs off on you. Again, how much you invest is up to you. There is no blueprint but one thing is certain: beauty is out there and it's there for everyone. Recreational fishing creates beauty. In prosaic terms this means:

- riverbank planting and bank stabilisation
- fish ways
- bushfire, drought and flood recovery initiatives, including emergency fish relocations
- monitoring river health and educating future generations
- waterway cleanups and sustainable fish education programmes
- shellfish reef restoration and shell recycling
- seagrass restoration
- mangrove and saltmarsh restoration
- habitat mapping
- kelp restoration
- citizen science including mapping, water quality testing, water bug monitoring
- resnagging
- wetland restoration

This is the sort of formula for beauty on the website of Ozfish Unlimited and likewise the website of Trout Unlimited – especially their projects pages – that warms the cockles of the heart of all lovers of beauty. Just in case this doesn't sound enough like an ad for them I had better be even clearer: Ozfish Unlimited and Trout Unlimited deserve your support and so does Fish & Game in New Zealand and so do all your local angling initiatives: all creators of beauty.

TV and reality

Switching on the TV means you fog your brain and pollute your soul. A newspaper, magazine or book leaves space for your own thoughts and images. The TV may do that for you too but you don't need to think about anything at all to switch on the TV except perhaps to remember where you left the remote. Going fishing on the other hand requires focus and concentration. Switching off the television and coming back from fishing are two entirely different experiences.

Let's ignore the TV couch potato and focus on the angler: as a rule the angler comes back from fishing happy, or at least relaxed and in a positive mood. Of course fishing can be frustrating at times. Usually you have yourself to blame, as in the case of the 'one that got away' because of a bad knot. Even so: there is no one else involved. It's all you.

The relaxation factor in fishing is central to the get-away-from-it-all myth. Fishing isn't escapism: you never get away from it all, especially not from yourself. If anything, fishing brings you back to yourself by virtue of the beauty experience, which is the source of joie de vivre. What happens when you go fishing is the opposite of what is claimed: when you go fishing you don't get away from it all, you come to terms with it all. 'It' means basically yourself, your world and the world at large.

In *Fisherman's Bounty* A.J. McClane shares this view with a slightly different emphasis and reasoning but the conclusion is the same:

Psychologists tell us that one reason we enjoy fishing is because it is an escape. This is meaningless. True, a man who works in the city wants to 'escape' to the country, but the clinical implication is that (no matter where a man lives), he seeks to avoid reality. This is as obtuse as the philosophical doctrine which holds that no reality exists outside the mind. Perhaps it's the farm boy in me, but I would apply Aristotelian logic – the chicken came before the egg because it is real and the egg is only potential. By the same reasoning the fluid content of a stream is nothing but water when it erupts from a city faucet, but given shores it becomes a river, and as a river it is perfectly capable of creating life, and therefore it is real. It is not a sewer, nor a conveyor of barges and lumber, although it can be pressed to those burdens and, indeed, as a living thing it can also be lost in its responsibilities. So if escapism is a reason for angling – then the escape is to reality.

Note the triad river, life and reality: the river creates life, life is real and so is beauty. Like hook, line and knot; beauty, life and reality are inseparable. You could actually push McClane's reasoning one step further: the one thing you can't escape from when going fishing is beauty. On the whole, though 'escape' should be dropped from fishing discourse because as McClane rightly points out, it's meaningless.

Talking of reality: winning is easy – anybody can win. Losing and failing are more demanding. Fishing is not a scoring sport like tennis or football nor is it a game. Catching a fish is not a victory. This is probably why you rarely see trophy shots with anglers holding the fish over their heads like a trophy cup. In fishing you don't win, you achieve. You can 'lose' a fish but even then you 'win' because you learn from the experience. The way to achievement is often paved with obstacles, failures, frustrations, disasters and disappointments, all of which the angler has to deal with. Not to mention, for example, getting soaked, eaten alive by midges, scratched by thorns and stung by nettles. It's quite amazing what anglers of all ages will put up with in the pursuit of fish.

Watching TV, playing chess or a computer game can be done in stages. You can record a programme, pause a video game and resume it at any time. While you can walk away at any stage from television or a game, this would make little sense in fishing. Of course you can stop fishing at any time. However, fishing can't be done piecemeal; you have to stay with it as long as you're fishing. There is another vital difference between fishing, TV and video games: watching TV or video games might be a complete waste of time whereas fishing time is never wasted time. 'Fishing' and 'waste of time' are mutually exclusive.

Float fishing ersatz

Time is of absolutely no importance to beauty. Beauty is timeless. The processes of life are just there and they don't 'need' time like we do. Time is a useful invention ticking away in all directions. For all practical purposes we think of it as linear and sequential. In this view of time, you catch your fish before you actually know you do. Suppose you watch your float and it starts bobbing. What you are seeing is already in the past because it takes time until you see it. These are infinitesimal fractions of time but fractions of time they are. The entire process of catching the fish happens in the past, so the moment you think 'now I have him', you already have him. Not that you need to worry about it: it's just the way it is.

When out fishing, time seems to fly or disappear. Nothing passes faster than a day's fishing or a fishing holiday. A snap of your finger and it's gone. The reason for this is that you're immersed in the timelessness of beauty. By fully focusing on fishing you have left linearity behind, opened your senses and perceived beauty (*remember the autopilot from page 133*). This line of reasoning resonates with Zen which contends that 'timelessness and spacelessness are the natural and original zone of all things including human beings, for they are all grounded in it.' Finding and staying in focus is another key theme of Zen.

Zen resonates when the poet and angler Ted Hughes explains that a fisherman stares for hundreds of hours at a float and that his very being rests on that float...

> ...fishing with a float is a sort of mental exercise in concentration on a small point, while at the same time letting your imagination work freely to collect everything that might concern that still point...

Taking this observation a step further it might not be unreasonable to understand Zen as a form of float fishing without actually going fishing – thereby saving the licence fee. If it wasn't for Zen, Zen practitioners would all be out float fishing. Contemplated from this angle, Zen then would be float fishing ersatz.

Nice to have: culture

In the wider sense 'culture' covers all angles of fishing. The hook is culture and so are the knot, the line, volunteer work (restoring rivers), how-to books, science, rod building and everything else. In the narrower sense 'culture' covers history, philosophy and all kinds of literature and art. The most important fact governing fishing culture is its universality. Fishing and fishing talk knows no barriers. The world over, the one that got away invariably is a big fish. Likewise that kind of story is universally met with scepticism. The prejudice that 'fishermen are born honest but they get over it' is so deeply engrained that the angler – if he wants to relate the truth – has no choice but to upsize the one that got away. Thus the angler factors in the downsizing that the listener is bound to make on the basis of that completely unfounded assumption that 'all fishermen are liars'. The opposite is the case: fishermen are master truth-tellers. Incidentally, if it is indeed true that all fishermen are liars consider the following: I am a fisherman and the last few lines are all blatant lies. True or False?

Be that as it may, fishing talk is universal: the weather, the tackle, the bait and so it goes on. Although this may seem to the nonangler the same over and over again, it is not. The variables in the fishing equation change on a daily basis which is why fishermen never run out of topics to discuss and they surely do not run out of challenges. Wherever you live it's the same themes.

The universality of fishing – recreational fishing – is underrated. What do I mean by that? Fishing is a human activity, an experience uniting people from different cultural and social backgrounds. Fishing knows no boundaries: if there ever was a true 'united nations' it's the 'united nations of anglers' because unlike real nations, anglers are united in peaceful purpose and culture. And not only since yesterday but since the dawn of time. This is the reason why recreational fishing should be declared a UNESCO World Cultural Heritage Activity.

We can only speculate what the fishing experience with hook and line was like 20,000 years ago when we believe the first fishing hooks were being used. Surely, just as today, there was pleasure involved in catching a fish for food and perhaps also the excitement of the chase. There might even have been a relaxing element in it because the angler was away from the hustle and bustle of the settlement. Coming back with a fish and the experience of catching it might have changed the mood of the angler and the atmosphere of the settlement. We don't know but it is not unreasonable to assume that fishing success was positively welcome.

As we get closer to the present there is more evidence. The Egyptian pyramids are a World Heritage Site and at their front door, so to speak, flows the mighty Nile. Kenneth J. Stein has researched the Egyptian experience and writes:

> It is easy to see that the Ancient Egyptians were the early innovators of modern day fishing and most likely, not any different from us.

The point to take home is that fishing has a long pedigree few other human activities have. That is one thing: the other is whatever the motive (subsistence, food, recreation) the experience of beauty was there from day one.

You don't need to know about beauty in order to experience it. It's just there, always was and it is waiting for you. Note that this applies to everybody and to everywhere.

Here is a glimpse into ancient China: in the article *Fishing expresses aspiration: spiritual freedom in ancient China*, the authors, Zheng Wei and Lu Yi, explain that getting close to nature and appreciating the beautiful natural scenery was an ideal choice for many scholar-officials to relieve the stresses of politics, indulging in their aspiration to lead a hermit life and pursue spiritual freedom.

> Putting on a straw rain cape and bamboo rain hat, the angler sits solitarily by the water, holding a fishing rod. This was a great relaxation to both his body and mind.

In the same article, the authors relate a tale in which an angler baited his hook with 50 oxen. He caught a leviathan and the tsunami which the fight created was felt for hundreds of kilometres inland.

Kiwis will, of course, immediately think of 'Te Ika-a-Māui' (*The fish of Māui*) the Māori name for the North Island. Māui is a hero of superhuman strength in Māori mythology and his fishing story probably outshines all others. One day he went out fishing with his brothers and hooked a huge fish but it wasn't a fish. He actually pulled up the North Island of New Zealand.

All human cultures and religions in one way or another have recognised water as the source of all life; and fish, if not fishing, as a key to life. This includes practical (food), metaphysical, poetical or mythical terms. Think of the Irish Salmon of Knowledge, the Chinese fairytale of the Princess and the Red Carp or of fish representations in sacred or profane art that you find everywhere and in all epochs of human history.

A study dated 1911 concludes:

>...the fish was held in awe in Eastern Asia as well as in Europe, in Egypt and in ancient Babylon. In prehistoric times it possessed a religious sanctity. It was a symbol of immortality as which it is found in different styles in graves, and it is freely used as an emblem of good luck.

We travel back to merry old England where in 1496 *The Treatyse of Fysshynge wyth an Angle* by Dame Juliana Berners was published. Some scholars doubt that the prioress of the Sopwell Nunnery near St. Albans was the author, but that's beside the point. There is, however, absolutely no doubt that the 'Treatyse' is a turning point in the history of fishing. Andrew Herd observes perspicaciously:

>The prologue of The Treatyse introduces fishing as a sport which is not merely equal, but superior to hunting and hawking, a sentiment that would have raised a few eyebrows in view of the rigid conventions of the time. The places of hunting and hawking were well established, but fishing was, by and large, a pot-filling exercise for the masses rather than a sport for the elite.

By elevating fishing to a socially acceptable fieldsport, Dame Juliana 'invented' English and European recreational fishing. Of course sport was not for everybody then, but over time fishing became affordable and the popular pastime it is today.

Beauty boosters

Going fishing is the quest for beauty; and nature conservation work is boosting beauty. You know 'ugly' when you see it. On a large scale you know it is ugly when rivers are burning from pollution like the Cuyahoga River (USA 1969) or the Schweizerhalle disaster in Switzerland (1986) when a warehouse fire released tons of deadly chemicals into the Rhine and literally exterminated all aquatic life. On a local scale you recognise 'ugly' by, for example, rubbish dumped into water. In between there are all sorts of combined ugliness such as pollution (not necessarily visible), invasive species (plants and animals) and unfriendly riparian agriculture. There are notable absences too, like the absence of insects, birds or any riparian vegetation to speak of. The absence of life is ugly too. Conservation is about replacing the ugly with the beautiful.

Terminology first: there is a difference between environmentalism and conservationism. Very broadly speaking, the environmentalist is philosophically, ideologically and politically motivated while the conservationist is more pragmatically focused. The dividing lines are blurred and there are overlapping areas but Jane Elder captures the essence of the distinction succinctly:

Environmentalist: The environment is to be saved, preserved, set aside, protected from human abuse.
Conservationist: The environment is something we use, so we have to conserve it and take care of it, so that others can use it in the future.
Environmentalist Stereotype: Greenpeace activist
Conservationist Stereotype: Local duck hunter

'Environmentalism' sails under many flags including deep ecology, bioregionalism, land ethics, eco-feminism, environmental-activism, animism, social ecology and eco-justice. They all, however, pivot around the same theme: the relationship between humans and nature.

Some scholars and politicians bend over backwards to forge unholy alliances between animal rights and environmentalism. It's easy really – as one activist put it: 'I believe in animal rights, human rights, land rights, water rights, air rights.' You can believe in anything you like as long as it is politically correct. It's the same political forces across the First World that embrace and push environmentalism and animal rights either separately or together. That is not surprising: after all, fundamental change in practically all areas of life is their agenda.

Whether that is brought about by environmentalism or animal rights or both doesn't really matter. In either case *1984* is already in the making. Exaggerated?

An indicator of the environmentalist zeitgeist is the Swiss Green Party parliamentarian Valentine Python who demands that 'climate sceptic remarks' must be a punishable criminal offence. Not that she is out on a limb on her own. Professor Richard Parncutt from the University of Graz in Austria goes a step further and would love to see the death penalty for climate sceptics. That is not just pure *1984*, it's the beginning of the end of Western democracies.

Environmentalism and animal rights are often brothers in arms and they certainly share a common enemy. As with animal rights philosophers probably the majority of environmentalists claim that Western culture (Christianity, capitalism, colonialism, liberalism, science etc.) is at the root of almost every problem that plagues the world. The classic statement of this view which is as popular as ever, is an article by Lynn White Jr published in 1967: *The Historical Roots of our Ecological Crisis*. White claims that:

> By destroying pagan animism, Christianity made it possible to exploit nature in a mood of indifference to the feelings of natural objects.

Note 'feelings'. He argues that modern science and technology are of Christian origination and together with the dogma of humankind's rightful mastery over nature brought about the ecological crisis. Further he says that Christianity 'bears a huge burden of guilt'. You just can't win, can you? Unless, of course, you implement the slogan: 'The optimum human population on earth is zero.'

It's interesting to note that Western environmentalism and animal rights always walked together – not hand in hand but side by side.

Here is the briefest of rundowns:

Bentham and Wordsworth

Jeremy Bentham (1748–1832) UK

Bentham's famous quote: *'The question is not, Can they reason?, nor Can they talk? but, Can they Suffer? Why should the law refuse its protection to any sensitive being?'* is the original inspiration of all animal rights as we know it today.

William Wordsworth (1770–1850) UK

William Wordsworth and other romantic poets like Samuel Taylor Coleridge (1772–1834) saw the effects of industrialisation. Wordsworth:

> *The world is too much with us; late and soon,*
> *Getting and spending, we lay waste our powers;*
> *Little we see in Nature that is ours;*
> *We have given our hearts away, a sordid boon!*

The romantics discovered the sublime and beauty in nature and celebrated it. They were the vanguard of modern environmentalism and conservationism.

Thoreau and Marsh

Henry David Thoreau (1817–1862) USA

Henry David Thoreau is often cited as being in the animal rights camp. After his Walden days he gave up fishing.

George Perkins Marsh (1801–1882) USA

George Perkins Marsh is sometimes called the father of environmentalism or the first American environmentalist.

Salt and Pinchot

Henry Salt (1851–1939) UK

Henry Salt's *Animals' Rights: Considered in Relation to Social Progress* is a milestone in the history of animal rights.

Gifford Pinchot (1865–1946) USA

Gifford Pinchot was a keen angler and the US conservation pioneer. He was the first chief of the US Forest Service.

Mentioning Gifford Pinchot and leaving out John Muir (1838–1914) is just not on. Not least because they illustrate perfectly my earlier point about the lines being blurred and the overlapping areas of environmentalism and conservationism. Lukas Keel writes in *Humanities*:

> The two have come to embody the conflicting philosophies at the heart of the American public land system: preservation vs. conservation. For Muir, nature was God, best preserved far from the degrading touch of man. For Pinchot, nature was a resource that ought to be sustainably shared among the most people possible. These opposing views might have made the two men natural enemies.

According to Keel, this didn't happen because both liked fishing. Another keen angler and a key figure of modern environmentalism and conservation was Aldo Leopold (1887–1948). His best known work is *A Sand County Almanac* in which he defines his land ethic as:

> A thing is right when it tends to preserve the integrity, stability and beauty of the biotic community.

Now there is food for thought. 'Stability' requires a little clarification. There is no such thing as stability in nature in the sense of being static. Nature is always on the move, imperceptibly slowly but moving nevertheless. The process of life creates relative balance so the different parts of an organism or a system can function properly. The stability of the biotic community as a whole would be the interplay of all the life processes in its given environment. In other words a healthy ecosystem.

The process of life produces changes, some of which might endanger the 'stability', like one species monopolising a resource or an invasive species upsetting integrity, stability and beauty. We humans are like all other parts of the biotic community; we have no special status. We are not really individuals but more like biotic units. If there are now too many human biotic units, threatening the integrity, stability and beauty of the biotic community, then some or all of us have to go. Humans would be managed like other wildlife and be shot, poisoned or whatever in line with the management goals of the biotic community. The follow-up question is, of course, who decides and on what basis? And if you pursue this you inevitably land in some authoritarian constructs – a sort of environmental *1984*. This is the reason why Leopold is accused of 'environmental fascism'.

If you think this is all a bit far-fetched think of the 'Project Tiger' campaign launched in India in 1973. The biotic community then lacked tigers. In order to restore tiger populations, tiger reserves were initiated. Fifty years later the Indian Prime Minister Narendra Modi celebrated the success as an achievement for the whole world. Not to be outdone, a spokesperson for the Global Tiger Forum stated that the success was 'unparalleled in conservation history.'

Even if I believe only a fraction of the information available on the internet the price was paid over decades by indigenous people. This is sad.

ABC News Australia reports (2023):

Several Indigenous groups say the conservation strategies, deeply influenced by American environmentalism, meant uprooting numerous communities that had lived in the forests for millennia. [...] the social costs of fortress conservation — where forest departments protect wildlife and prevent local communities from entering forest regions — is high.

Sharachchandra Lele, of the Bengaluru-based Ashoka Trust for Research in Ecology and the Environment, said the conservation model is outdated. 'There are already successful examples of forests managed by local communities in collaboration with government officials, and tiger numbers have actually increased even while people have benefited in these regions,' he said.

Vidya Athreya, the director of Wildlife Conservation Society in India, who has been studying the interactions between large cats and humans for the last two decades, agreed.

'Traditionally we always put wildlife over people,' Ms Athreya said, adding that engaging with communities is the way forward for protecting wildlife in India.

Mr Shivu, from the Jenu Kuruba tribe, wants to go back to a life where Indigenous communities and tigers lived together.

'Wildlife over people': replace wildlife with biotic community and you have Leopold's problem in a nutshell and in real life. Wilderness is where wildlife happens, but there is no such thing as 'wilderness' in the sense of not touched or tampered with by humans. There are remote and wild areas for instance in Siberia or the Amazon, but directly or indirectly (e.g. via air pollution) the human presence can be detected everywhere. Earth by and large is a colourful collection of cultural landscapes: over the millennia humankind has shaped, not to say designed, the environment. This is true of all cultures. In order to survive, humans everywhere had and have to adapt and interact with the environment.

The epitome of 'design' are national parks (i.e. conservation areas). They are truly cultural achievements. There are perfectly good reasons to limit access to certain areas; but to exclude people entirely as they do in 'fortress conservation' or to impose conservation top-down is generally not a good idea. Western NGOs dictating and enforcing conservation on local people outside Europe or North America is the same kind of neo-colonialist mindset as can be found in animal rights and their iron fist in a velvet glove approach to indigenous people (*see page 102*).

Back to Aldo Leopold: I am not sure whether the eco-fascism charge is rock solid. If you take his statement that 'a thing is right when it tends to preserve the integrity, stability and beauty of the biotic community' and add another observation of his that 'we can only be ethical in relation to something we can see, understand, feel, love or otherwise have faith in' then the picture changes.

We humans are the only members of the biotic community who can relate to the environment on those terms. We relate to the reality around us as individuals, not as a collective entity. Understanding is an individual learning curve and not a dictate from above. Interacting with the people and the world around you on the basis of understanding (seeing, feeling, loving, faith) and the corresponding reciprocity sounds more like a human-friendly approach rather than a fascist dictate. Be that as it may, integrity, stability and beauty are worth aiming for, whether in terms of preservation or in terms of conservation or some middle ground.

All that history and all these seemingly remote debates and hybridisation between animal rights and environmentalism are real when it comes down to the all-important area of defining and making law. The underlying issues are the same everywhere and so are the controversies and the political battles.

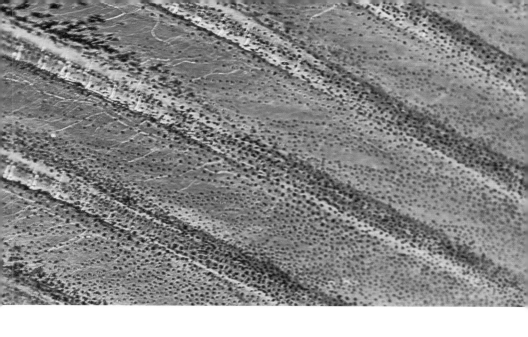

Chapter 7
The End Is Nigh

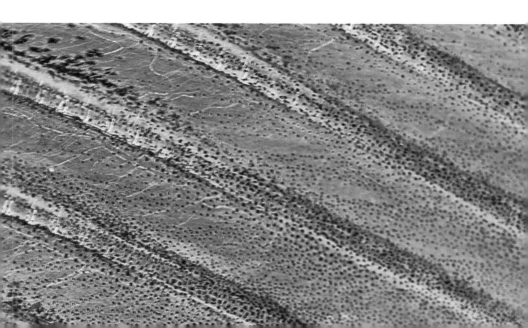

Miracles do happen. Since about 1960, time has miraculously stood still at five to midnight. Smog, DDT, nuclear war, the new Ice Age, acid rain, the population explosion, AIDS, bird flu, desertification, the ozone layer, dying forests, global warming and many minor apocalyptic forebodings point to the end of time. The red lists are getting redder and redder, and is there any furry, heartbreakingly cuddly wild animal for which it is not the eleventh hour? We are beset by all kinds of potential major and minor disasters and it's all your fault (white, First World, wrong gender) and you're meant to feel guilty about it. But do not despair: whatever happens you get off the hook by donating to the right cause. The historic predecessor to fundraising for worthy causes was the sale of indulgences. The remarkable thing is that there seems to be a strange equation at work: the problems grow in proportion to the funds raised.

Big problems are big business. Like their commercial counterparts, all the major environmental NGOs are global players, or aim to be. That is actually a good thing as it keeps politicians on their environmental toes in a way an individual never could. On the other hand, the fact that this is necessary speaks volumes about the conservation achievements of the political establishment. Yet another take on the situation with governments and multinational NGOs could be to say that a bit of ongoing crisis is of mutual benefit: *honi soit qui mal y pense (shamed be whoever thinks ill of it)*.

In terms of effectiveness, the balance sheet of the big players can also raise a question mark or two. From a brutally utilitarian point of view, saving a few charismatic species around the world is, in terms of biodiversity, probably less effective than urban or suburban wildlife gardens.

Why? If you succeeded in converting only half of the world's suburban gardens into wildlife gardens, the impact would be huge. But alas pesky insects, busy beetles, mighty moths, nondescript grasses, stinky mushrooms and noisy birds are hardly as spectacular and adorable as a bamboo-munching bear. On top of that, most of the good you did would be invisible (e.g. millions of little critters and organisms in a healthy soil). But there is a reward: beauty. The beauty of wildlife gardens has its price. It requires time, money and know-how.

Think globally, act locally?

This slogan was allegedly coined by environmentalists in the 1960s. It reflects perfectly the paternalistic and centralist mindset that has always been part of environmentalism and animal rights. The big thinking is done in big cities by big heads, and the pea-brained yokels then carry out the small tasks.

The big players have been extremely successful, and it would be a real disaster if, for example, the WWF collapsed for some reason. Unthinkable, but so was the coming down of the Berlin Wall. The WWF is like a big bank: it's too big to fail! Some NGOs make themselves indispensable in such a way, that they become at least partly funded by the taxpayer (government subsidies). Thus, indirectly, I end up paying towards organisations that are clearly anti-fishing and anti-hunting.

The UN is another global player driving environmental causes. As I write, today is 22 April 2023 – International Mother Earth Day. In a more than slightly bizarre message, António Guterres (UN Secretary-General) informs humankind that we (human beings) seem 'hellbent on destruction' and are engaged in wars on nature:

We must end these relentless and senseless wars on nature.

That is an astounding statement. Who are 'we'? It certainly doesn't include the indigenous people from whom – according to Guterres – 'we must learn'. Does he mean all the other human beings? Maybe he does, because he says 'we' have to help 'communities who have done the least to cause the crisis.' Note 'the least' to cause a crisis: so the poor are guilty too... at least a little bit. Does 'we' include all the millions of people around the world who are seriously trying to make a change?

Am I at war with Mother Earth? I wasn't aware that I was but now Guterres is telling me and everybody else, it surely must be true...

Never mind – the UN knows where and how peace is to be found. Guterres moves on to say:

At every step, governments must lead the way.

Sure! They have led the way in the last 70 years and what's the result? 'We' are at war with Mother Earth, that's what. A definition of madness is doing more of the same and expecting a different result. Besides, what government did he have in mind? There are all sorts out there. And what do you make of the complete absence of Father Sky?

Not to be outdone, David Cooper, Acting Executive Secretary of the 'Convention of Biological Diversity' tells us:

Today, this International Mother Earth Day, we reflect on our relationship with and our dependence on nature.

Biodiversity is the heartbeat of Mother Earth – the variety of life on our planet. It is the air we breathe, the water we drink, the soil we walk on, the food we eat, the trees we seek shade under; it is everything. Unfortunately, that heartbeat is becoming faint. Biodiversity is being lost at unprecedented rates. That loss is being accelerated by climate change, and many other human factors, and our future is being threatened. It is time to resuscitate Mother Earth's heartbeat and work towards the global vision of 'Living in harmony with nature' by 2050.

Really? Doom, gloom, guilt. There is also the question: Who is Mother Earth? Guterres and Cooper talk as if it was self-evident that everybody knows her. Uncle Google provides a range of profiles, but the *Encyclopedia Britannica* offers a concise summary:

Earth Mother, in ancient and modern nonliterate religions, is an eternally fruitful source of everything. Unlike the variety of female fertility deities called mother goddesses, the Earth Mother is not a specific source of vitality who must periodically undergo sexual intercourse. She is simply the mother; there is nothing separate from her. All things come from her, return to her, and are her.

Assuming that Guterres and Cooper had this in mind, are they endorsing a kind of earth religion? Is Mother Earth heralding a shift away from science-based environmentalism to a faith-based approach? Or a marriage of the two? Whichever it is, the ride is probably going to be even bumpier than it already is.

It's difficult to make sense of what Guterres and Cooper are saying, but after all, we are in global politics here. Since the 1992 signing of the Rio Declaration (the famous Earth Summit) the contribution to biodiversity by those participating seem to be basically words, conferences, protocols and summits. I have lost count of the summits and the solemn declarations. If only a fraction of all the hours and the money involved (31 years, thousands and thousands of participants, advisers, back office staff, equipment, subsidies, research, air, sea and land travel, etc.) had been directed to some meaningful and direct contribution, like habitat restoration, anywhere specific in the world, biodiversity would have been the winner. With the total of resources spent since 1992 on a load of words, you could have probably bought half of the remaining Amazon Basin, and protected it.

Think now of the hours and money your fishing or hunting club has put into biodiversity in the last 31 years. Look at the results: you have beautified the world and made it a better place. You haven't led a war and you weren't 'hell-bent on destruction' as Guterres has it. It looks more like Guterres is hell-bent on stifling the goodwill there is by lumping together blame and guilt allocation ('we') and throwing it indiscriminately at everybody, including all those engaged in meaningful conservation work (volunteering). It is an insult to good people, goodwill and good work. If you look at the balance sheet factoring in the ecological footprint of Rio 1992 and its follow-up operations (remember all that climate-saving jet travel), the bottom line will be that your fishing or hunting club is the clear biodiversity winner over Rio & Co. Yet there are millions of anglers the world over accused of being 'hell-bent on destruction.'

Most of us have to make an honest living outside summits and general assemblies and don't find the time to engage with environmental causes as much as we would like. Rather than viewing donations to environmental NGOs as indulgences you could see them as a mandate and endorsement. Your donation enables the professionals to do a good job and pursue projects that contribute to biodiversity. Being hypercritical of global NGOs doesn't mean ignoring or denigrating the good work they do. At least some of them have got a vital role to play, and for exactly that reason keeping an eye on them and an open mind seems to be a good approach. If you decide to donate to an environmental NGO, you then are spoiled for choice. There is a lot of competition out there. If in doubt, opt for a fishing or hunting organisation. But whatever, think local, act local, be global.

What did the fish say when it hit the wall?

Dam!

Conservation efforts by anglers and nonanglers might be belittled as hopeless romanticism because you can't turn the clock back. It is certainly true that a presumed ideal state in the preindustrial past can't be resurrected or restored, because it is in the past. However, why look for beauty in the past if a beautiful future is to be had? Turn the clock forward and catch up with the past.

Dam removals are an absolutely fascinating subject and some of the success stories are breathtaking. The Elwha dam removal in the United States is such a story and so are the 239 dams and weirs removed by Dam Removal Europe in 2022. By letting streams and rivers flow, they revitalise the environment: they bring back life and beauty.

There are thousands and thousands of restoration projects driven by a fishing interest and it seems unfair to single one such project out. But The River Teign Restoration Project is representative of the spirit and reasoning of many such projects big and small:

> Rivers have always been important to humans wherever they are in the world and the River Teign is no exception. As a landscape feature, it is a waymarker, a recognisable and passable route through the landscape to be followed on foot and where the water conditions allow as a means of transportation and communication, particularly in the lower reaches. As a life-giver water is not only a resource in its own right but it also provides habitats for many other species of animal and plant which in turn have provided us with food. The river has always provided sustenance and livelihoods. Fish are an obvious and important part of this story and there is a rich history of fishing from the netsmen of the estuary to subsistence fishing along its length. In more recent times the sport of angling for salmon and trout has been a popular pursuit and it is via this activity that many have got to know and fall in love with the river and the communities within its vicinity.

'Wherever they are in the world'. *Yes: think local, act local, be global!* If you look at invasive species from an untrammelled global viewpoint, there are no invasive species. If you're a true globalist, invasion is migration and you understand all natural and human history as a story of migration. If you're a local globalist, invasive means a plant, animal or micro-organism that has invaded your local ecosystem and threatens the integrity of it. Invasion – be it human-aided or self-introduced – always has negative impacts. A few random examples of invasive species are possums in New Zealand, carp and pike in Spain and the hippo in Colombia. On the other hand, an alien species can be a friendly newcomer to your local ecosystem. Crops may be alien species, whether apple trees or trout. Invasive or alien, it just depends where your global local is.

Boosting beauty has its pitfalls – not that beauty itself is the issue but some aspect of it. Remember Masaryk and the trout:

Wherever the trout are, it's beautiful.

Conservationists would say 'it all depends...' The problem originates in the late nineteenth century when, as Andrew Herd puts it:

...the oceans were criss-crossed with the wakes of ships playing an epic game of backgammon with our trout populations and so few records were kept that it is hard to keep track of what went where, save to say that the genetic pool became irretrievably mixed up.

A particularly intricate and puzzling situation arose and persists in North America. The online *Smithsonian Magazine* reported in 2007:

Whether a trout is a nuisance or a valued member of the community depends upon where you stand on the map. Of the four major trout species in the United States — rainbow, brook, cutthroat and brown — only the brown trout was introduced from abroad, but any of the four might be considered invasive when introduced into a new watershed. Thus, a rainbow trout (*Oncorhynchus mykiss*) transplanted from its native California to Virginia is regarded as a non-native in its new home; by the same reasoning, an Eastern brook trout becomes a pest in Western streams. It has displaced resident trout from the small rivers and lakes of Montana, Colorado, New Mexico and other mountain states. The brook trout's main victim is the cutthroat, so called for the bright slash of crimson under its jaw. Squeezed on one side by invasive brook trout, native cutthroats are also under challenge from rainbow trout, a cousin introduced from the Pacific Coast.

In order to restore the natural state – reintroducing the proper indigenous species in the right river and streams – piscicides are used to kill the alien species. Once they are exterminated, the local species are reintroduced. Poisoning fish sends shivers down anybody's spine and makes you think twice. Nevertheless, the evidence in the scientific literature seems to suggest that poisoning non-native species and reintroducing natives, works. There are question marks left, however. One study (2021) puts it like this:

> Non-native fish eradication via the piscicide rotenone is an effective tool for fisheries management and conservation of native species. However, the long-term effects on nontarget organisms, including benthic invertebrates and zooplankton in alpine lakes, are understudied and are poorly understood.

To Kiwis 1080 immediately springs to mind. For non-Kiwis, 1080 (pronounced ten eighty) is a poison that is extensively used in New Zealand to fight possums and other predators. It is said to kill 95% of possums and 100% of rats and other predators in a given area and claimed to be otherwise largely harmless (biodegradable). In the vision of achieving a predator-free New Zealand by 2050, 1080 plays a key role. The application of 1080 is backed up by serious science and its success seems obvious. However, it is hardly surprising that there is opposition. Poison is intuitively wrong. Anybody in their right mind will spontaneously be concerned about its use and effects on drinking water, on livestock, game, birds and all other kinds of wildlife. The Department of Conservation (DOC) is in charge of the programme and tries to dispel doubts by providing in-depth information. DOC documents the use, research and progress made with 1080 in great detail. But who cares about rational pros and cons – that's really yesterday's news; completely out of touch with reality. Unfortunately the Royal New Zealand Society for the Prevention of Cruelty to Animals (SPCA) pulls us straight back to the heart of ugliness:

> SPCA is against the use of poisons to kill animals due to the level of suffering they cause, as well as the nature of their use. We would

like to see a ban on the use of poisons such as 1080, because these substances cause such intense and prolonged suffering to animals that we believe their use can never be justified.

Never! This is applied animal rights ideology. Pure as the Manifesto (*page 124*). Forest & Bird is one of the New Zealand organisations engaged in the fantastic project Predator Free 2050. Its message to the SPCA is as unambiguous and incisive as you can get it:

Without scientific, ethical, and precision pest control, of which 1080 is a key tool, there is no way to protect our native animals from the overwhelming numbers of introduced predators. Giving up 1080 would lead to an ecocide of huge proportions in New Zealand, and the SPCA need to understand this is the outcome of their pest control position.

I told you that animal rights are conservation wrongs. The SPCA's position on 1080 is an instance of this.

Back to fishing: beauty easily comes to you when fishing, the hard bit is boosting and caring for it. There is more to fishing than catching fish. In fact it's a damn sight more demanding than you would think.

Natural stupidity and artificial intelligence

Of course I am biased because I love the beauty that is in fishing. Of course the conclusions of Cholmondeley-Pennell, Rose and Key suit me better than any other. Of course the powers of my intellect are limited, of course my sources are bias-selected, of course I am anthropocentric, of course I create strawmen, of course I am inconsistent, and of course I might have got it all wrong (after all I am white and male). But why bank on natural stupidity and narrowness when boundless artificial intelligence is but a few clicks away.

Here is what AI (scite.ai) came up with when asked whether fish feel pain:

The evidence for whether fish feel pain is still a topic of debate in the scientific community. Some studies suggest that fish are capable of nociception, the simple detection of a noxious, potentially painful stimulus and the reflex response to this. However, distinguishing between nociception and conscious pain is difficult, and claims that fish feel pain remain unsubstantiated.

Recent studies have shown that fish display aversive behavioural and physiological reactions and a suspension of normal behaviour in response to noxious stimuli that cause pain in other animals and humans. In addition to these behavioural responses, scientists have identified a peripheral nociceptive system and recorded specific changes in the brain activity of fish during noxious stimulation. Therefore, teleost fish are now considered capable of nociception and, in some opinions, pain perception. Pain may be species-specific, and different forms of pain may have different behavioural responses. However, some studies have concluded that fish do not have the neural apparatus to detect pain. The debate on whether fish feel pain is ongoing, and further studies are needed to investigate the responses of fish to noxious stimuli.

The same AI on the question 'Is fishing cruel?'

In conclusion, while there is no clear consensus on whether fishing is cruel, research suggests that human activities, including fishing, can affect the welfare of fish. Catch-and-release angling can reduce injury, mortality, and sub-lethal alterations in behaviour and physiology, but post-slipping mortality rates can be unacceptably high. Consumers have positive perceptions regarding sustainable fish products and show a willingness to pay a premium price, but fish welfare has recently become an important issue in the EU. Finally, scientific evidence suggests that fish do indeed feel pain and suffer.

I wonder about AI – could it be that, just like me, it has a bias? So I asked, 'Is AI biased in moral questions like cruelty to fish?' and it answered:

> The provided research results do not directly answer the question of whether AI is biased in moral questions like cruelty to fish. However, the results do suggest that ethical considerations and mitigation strategies should be embedded in the design, training, and deployment of AI systems to ensure social good while still benefiting from the huge potential of AI.
>
> There are concerns about potential bias and discrimination in AI systems, and algorithmic bias and discrimination have been identified as ethical issues related to AI in healthcare. Additionally, the lack of transparency regarding training methods and insufficient validation techniques for assessing performance accuracy are reproducibility issues that can impact the use of AI in healthcare decision-making. Therefore, it is important to ensure that AI systems are aligned with human values and goals, and that they are explainable to build trust and understanding in AI decision-making.

The last sentence could be straight from *1984*. Trust AI and its 'aligned goals'? No way. Henry Cholmondeley-Pennell concluded his thoughts about the ethics of angling in regard to cruelty, that you can enjoy it 'without compunction'. On that specific question and in his time, that's it, yes. However, time and the world have moved on and today the ethics of recreational fishing can't be seen just in terms of pain and cruelty. The scope of the ethics of recreational angling has widened considerably.

You can't make an omelette...

An anaesthetic numbs some or all of your senses. 'Aesthetic' is Greek in origin and means 'of or for perception by the senses, perceptive'. Since the nineteenth century it has come to mean 'of or pertaining to appreciation of the beautiful'.

Fishing is the quest for beauty and thus an aesthetic experience. Of course that experience can be had by hikers or simply by sitting on a bench and enjoying a beautiful view. Kayakers, windsurfers and all kinds of aquatic sports produce a kind of aesthetic experience but a lot of attention goes into the handling of the equipment and delivering performance. Fishing and hunting draw you into the works of nature in a way no other sport does. The hiker and the kayaker are using nature as a facility whereas the angler actively participates in the processes of nature, which is why the aesthetic experience, the experience of beauty in fishing and hunting is unique.

Our animal rights friends wouldn't agree with any of that. They point to the suffering fish and to you, the bloodthirsty fish killer. What could possibly be beautiful about catching and killing a fish, they ask? The very question reveals their monochrome world, their tunnel vision and their oppressive urges. They are incapable of understanding that fishing (hunting) is not an act but a process in which catching and killing is but an episode. It is beyond their horizon to see nature as a positive given that includes us. We are an integral part of nature (unless we opt for the supernatural or extraterrestrial option I mentioned earlier in which we are not part of nature).

The beauty of fishing relates to the integrity of the experience and not to any particular action or set of actions in it. Going fishing doesn't start when you go out of the front door and nor does it stop when you come back and close the door behind you. It starts before that, with preparation – and ends with cooking and eating. And in fact going fishing started way before the preparation and doesn't really end after dinner. You are already planning your next fishing trip and remembering your last one.

Look at the perch: its yellowish olive body with the vertical bands is perfectly streamlined. Its flaming gold and orange fins are of incredible depth and radiance. What a breathtakingly beautiful fish. Look at the pike, the roach, the trout, the salmon, the tench, the rig, the gurnard; or at any of the stunning fish in the coral reef. They are absolutely awesome creatures, full of aesthetic appeal, just like birds, which is why snorkelling and birdwatching are such addictive pursuits. Aesthetic admiration of fish is an argument against all kinds of fishing which I fully understand and accept.

Fish are too beautiful to be caught and killed. But then you would have to step back from practically everything you eat. Think of the perfect proportions of a single stalk of wheat or the elegance and sheer beauty of a roe deer. You would have to become an aesthetic ascetic, and you would have my highest respect for that. But on the whole, the overarching argument for me would be that you can't make an omelette without breaking the eggs.

Buy your licence and abide by the rules!

That's it. It's the ethics of angling in one sentence. It's a minimalistic approach but perfectly OK. There is no need to run the entire obstacle course as I do in this book. Unless you want to, of course. The licence of a modern and science-based fishery (private or state) that is well managed in a forward-looking way includes all the considerations about fish welfare and conservation you need. Its rules are there for a reason and not there to be broken. Abide by the rules and you're fine. The licence fee contributes to the running costs and also to conservation; nevertheless you are of course free to engage beyond the bare minimum.

There is only one truth and one law

If the last 160,000 years are anything to go by, it seems to be part of human nature that we are fallible. We might, however, be evolving in the right direction because some of us, predominantly animal rights philosophers and their acolytes, are apparently already there. They have it all sussed out and there is only one indisputable, immovable and eternal truth and law: theirs. Unfortunately, the rest of us are too ignorant to recognise this.

All philosophies (in the widest sense of that word) involve prejudice, politics, contradictions, beliefs, inconsistencies, uncertainties, errors, fantasies and wishful thinking. In the real world, practical everyday morality is a middle-of-the-road process. Some parts of right and wrong are written in law and seemingly here forever; smaller chunks are there in the open, where the individual makes informed choices. The practice of fishing and all other fish uses are largely based on informed choice. You can evaluate the facts, weigh up the pros and cons and make an informed decision.

The 'infallibles' want to change that. They want their truth cemented in laws so you have to abide by their rules – reel in and forget about fish fingers.

The animals rights' law theorists, however, have to tackle two main problems:

1. Peter Singer's pain-centred consequentialism (sentience) and Tom Regan's rights-centred nonconsequentialism (intrinsic value) are mutually exclusive.

2. Both animal liberation and animal rights are incompatible with environmentalism and conservation. 'Species' are not individuals, neither are 'ecosystems'. They can neither feel pain nor have intrinsic value because they are not persons or subjects of a life.

The way out is easy. **The law can accommodate anything regardless of sense or nonsense. Anything can be law. The law can declare the world pretzel-shaped and pretzel-shaped it is.** Far worse things have been accommodated under the heading of law than the charming idea of a pretzel-shaped world. In fact the law is the animal rights movement's most valuable ally. It can mould together the most contradicting and ill-informed concepts into binding obligations for 'animal welfare' action.

Your fishing boat is a battlecruiser

Saskia Stucki is a highly acclaimed scholar. The animal rights world is at her feet. No surprise, because she manages the squaring of the circle – the fusion of mutually exclusive philosophical ideas in legal terms. As if touched with a magic wand (not a fishing rod), all contradictions disappear. Stucki's article *Beyond Animal Warfare Law* is, depending on your point of view, a most remarkable scholarly achievement or plain bull. The central contention is:

> Based on a comparative analysis with the law of war, this article argues that animal *welfare* law is best understood as a kind of

warfare law which regulates violent activities within an ongoing 'war on animals,' and needs to be complemented by a jus contra bellum [law against war] and peacetime animal rights.

She then juxtaposes a battle description (*Solferino*, 1859) by Henri Dunant (founder of the Red Cross) with a slaughterhouse scene by Upton Sinclair (*The Jungle*, 1906) suggesting one equals the other.

Stucki's contribution is the attempt to align and *level* animal welfare law with international humanitarian law. Stucki sketches her captivating argument as follows.

This article puts forward a novel analogy between *animal welfare law* (AWL) – the body of law governing the protection, and alleviating the suffering, of animals caught in situations of exploitative use – and *international humanitarian law* (IHL) – the body of law governing the protection, and alleviating the suffering, of humans caught in situations of war and other armed conflict. It likens the *humane* impetus informing AWL in its attempt to humanize innately violent and inhumane practices of exploitative animal use to the *humanitarian* thrust undergirding IHL in its endeavor to humanize innately violent and inhumane practices of warfare.

In this light your fishing boat turns into a battle cruiser because of your innately violent and inhumane angling intent:

1. Battlefields
2. Slaughterhouses
3. Fishing boats (recreational fishing)

Battlefield equals slaughterhouse. Slaughterhouse equals fishing boats and looking at it from the fishing boats side: fishing boats equal slaughterhouses and battlefields. In no time you're back to the animal rights fundamentalist stance: 'A rat is a pig is a dog is a boy.'

Existing animal welfare legislation is not welfare legislation but the rules that govern our war against animals. The war against animals is on now. Stucki:

> ...I submit that animal *welfare* law is best understood as a kind of *warfare* law that governs violent activities within the war on animals.

This means that the farmer down the road is at war with his sheep and cattle. It would also mean that you are at war with your dog or any other pet you love and care for: your dog is a prisoner of war that you exploit for your own ends.

War. According to Guterres (*see above*) we humans are at war with Mother Earth and Stucki has us specifically at war with animals. War, however, is an intrahuman fact and requires at least two parties. Nature and animals don't know anything about 'war'. There are no 'combatants' either in nature or the animal kingdom. Furthermore there aren't too many people around who would see themselves as 'combatants' intent on destruction and subjugation of nature and animals. Consider just a few groups:

- Surely the staff, sponsors and donors of all the environmental organisations don't see themselves as engaged in a war against nature and animals.
- Surely all the volunteer workers (not just anglers) involved in conservation projects don't see themselves as warriors against nature and animals.
- Surely all the animal rights people don't see themselves as campaigners against animals.
- Surely all the people who love and care for their pets don't see themselves as on the warpath against nature and animals.
- Surely all the drivers who slam on their brakes in order not to run over an animal are benevolent.

The long and short of it is: very few people – if any – see themselves as at war with nature and animals. The blanket indictment of all the decent people by Guterres and Stucki is not just insulting, it is absurd and cheap.

Nevertheless: who then do Guterres and Stucki have in mind with their war and their talk of 'we'? I don't know and can only speculate.

Imagine, for example, a single mum with three kids and a dog. Surely she must be at 'war' because she keeps a prisoner of war (dog), has added to overpopulation (three kids) and runs a car (fossil fuel)? That's absurd. Of course she is not at war as she bravely soldiers on in the daily grind making a living and a future for her kids. Like most people she wouldn't dream of participating in a war against Mother Earth or animals. Do Guterres and Stucki see themselves as combatants? Hardly. Does that actually leave anybody at war with Mother Earth and animals? Let's refocus on Stucki. However brilliantly she argues she can't argue away the uniqueness of humanity. In fact Stucki applies humanness to bring animals – at least some of them – to the level of what they are uniquely not: human.

Equating human life with animal life leads to some very ugly conclusions. PETA's 'Holocaust on your Plate' campaign is a case in point. In this campaign, genocide was presented as the moral equivalent to eating meat. The billboards juxtaposed, for example, baby pigs and concentration camp children prisoners with the headline 'Baby butchers'. You wouldn't expect anything less from the nice people of PETA. Noblesse oblige.

They don't walk alone. Some really nice people from the political establishment and, of course, university teachers, follow the same line. One Austrian professor has it that we (those thinking the holocaust comparison is absolutely barbaric, stupid and degraded) are not 'mature' enough to see the truth of the comparison. The website where he argued this has been removed. It is notable that most people publicly endorsing or slithering around the holocaust comparison have a university degree. For the most part, the rest of us, with our common sense still in working order, intuitively grasp the monstrosity of the comparison. It's tempting to observe that the goal of higher education seems to be to drive common sense out of people. This remark surely qualifies as anti-intellectualism which, of course, it is not – but that's another story for another day.

Stucki's war analogy is so wrong for yet another reason: it denigrates the achievements and goodwill of all the people actively engaged in animal welfare. The pioneers of animal welfare hardly saw their default position as being at war against animals. Nor would they be particularly pleased to learn that their efforts merely humanised warfare against animals.

Stucki aims, like the Manifesto, for the end of all animal use and couches this in the following terms:

> Therefore, in order to prevent the war on animals, rather than merely humanize warfare against animals, we need to look beyond existing animal warfare law.

'Prevent' means fully-blown legal animal rights which would mean the end of all animal use i.e. the end of war except of course for those uses which are deemed to be 'necessary' by the lawmakers i.e. law dictators. Obstacles or blatant absurdities and contradictions can easily be overcome by 'new conceptual vocabulary'. I spare you a discussion of the new vocabulary. I can't help remarking though that 'new conceptual vocabulary' is the kind of idea leading to Newspeak in Oceania (*1984*).

Stucki does not mention recreational fishing or fish. Nevertheless there is no reason not to extend her analogy to all the small fishing boats I see when I look out of my window onto Tasman Bay right now. Out there all these men and women are fighting a war against fish. It is 'collective violence' done to sentient beings, causing unnecessary suffering. 'Unnecessary suffering' is the magic key because those who can define unnecessary suffering and enforce observance of this, hold the power over your fishing. And a lot else. Or in more warlike terms: legal animal rights grow out of the barrel of a gun. Once you control the big guns, you define and alter the playing field and shift the goalposts however and whenever it suits you.

Judging by the academic company she keeps, Stucki wouldn't be too keen on recreational fishing. In her article she acknowledges the comments by Dinesh Wadiwel, an Australian social scientist.

He concludes an article about recreational fishing in Australia:

> ...one strategy for animal advocates to explore is to offer alternatives to fishing as recreation. Providing ways to offer pleasurable alternatives to hunting, injuring and killing animals 'for fun' of course provides a direct route towards challenging the epistemic dimension of human violence towards animals. Providing people alternatives might be one way to challenge this ongoing and pointless hostility.

The 'pointless hostility' is challenged and fought everywhere and by all means: PETA Germany demanded that a village called Fischen (fishing) change its name to 'Wandern' (walking) because Fischen (fishing) is a cruel pastime. Not entirely pointless: it got nationwide media attention there.

Nothing PETA does is pointless. It is in fact very, very clever. An annual business turnover figure of USD 82,226,083 (2022) is no joke. There is also mention of 9 million members and supporters in the 2022 annual report. This means you have at least 9 million people pointing their moral finger at you when you head to fish. If you're not convinced that animal rights is real, have a look at the annual reports of organisations in your neck of the woods that are campaigning for animal rights and those leaning towards animal rights. Add to those the public and private funds that go into research motivated by animal rights and you will see it's a veritable industry. It's big business.

The network of the philosophical and legal elite is global. Animal rights is no longer confined to the ivory towers. The philosopher kings and the legal eagles are out there in politics trying to steer us to a better, cruelty-free world – namely to scrap the battle cruisers. They are on a mission, and probably most of them sincerely believe in their cause. They are intelligent, young, well-funded and want to change the world. They don't appear troubled by the misanthropic and pessimistic basis of their plans for a bright future.

These plans could be described as instances of complete rubbish. Harry Frankfurt observes: 'One of the most salient features of our culture is that there is so much bullshit.' Bullshitting is not lying but an attitude which can't be bothered with facts or truth. I may be guilty of it myself but have no worries accepting that I might get overenthusiastic about this, that or the other because I believe there are facts and that logic and common sense will put me right. On the other hand you would have a hard time even to bring facts let alone common sense onto the radar of animal rights philosophers, activists, scientists and lawyers. Their bullshit and dogmas – FFP being one of them – override any other consideration. This matches perfectly with the vegan mum who refuses to free her child from head lice in order to respect the intrinsic worth and sentience of these animals.

Because it includes fish, here is a final example of how legal reasoning can turn water into wine.

In the article *Sentience and Intrinsic Worth as a Pluralist Foundation for Fundamental Animal Rights,* the author Jane Kotzmann, explains:

Sentience and intrinsic worth as a conceptual underpinning for animal rights hold clear benefits in that

(i)   the concepts are already embedded in many legal systems,

(ii) sentience would enable the development of animal rights to be built on the established interest theory of rights, and

(iii) sentience directly links to the justification of rights as being primarily concerned with the prevention of pain and suffering.

Great. The shortcomings of one concept (Peter Singer: pain, sentience, interests etc.) are patched up with the weaknesses of the other (Tom Regan: intrinsic value, subject of a life etc.) and vice versa. You can have your animal rights cake and eat it. You actually can! Who cares that in detail it doesn't bear scrutiny? It's all a bit vague but that is the lawyers' future bread and butter. Where does it lead?

To consensus.

Kotzmann:

> Like the concept of human dignity, which plays a foundational role in human rights law, the concepts of animal sentience and intrinsic worth have the benefit of bringing people to a point of consensus.

And, as you probably suspected, the consensus about fish, say the antis, is already well established:

> It is generally agreed that all vertebrates are sentient, and studies in relation to 'amphibians, reptiles, fish, cephalopods and decapod crustaceans' indicate that they are also sentient.

Stick to the dogma and never mind the facts because there is more at stake here. As Jane Kotzmann puts it:

> As it stands, the world is an unjust place in which human beings disrespect and exploit other sentient beings and our shared environment. To transition towards a world in which human beings can live in harmony with other animals and the environment with which we are interdependent, deep cultural change is required. Human beings must learn to recognise the intrinsic value of other animals and the environment and develop the law to ensure that such value is respected and protected.

Must we? *We have ways of making you talk...*

This brings us full circle back to the central question of FFP. As outlined earlier it is certainly not 'generally agreed' that fish feel pain. The FFP dogma in conjunction with animal rights is also extremely political and carries a massive economic impact. Globally. The implications of a full implementation of animal rights ideas range from promoting politically correct science, dismantling liberalism and enforcing draconian *1984* style measures (cancelling culture, language control etc.) through to cultural imperialism and an economic fallout of incredible dimensions. The FFP debate affects *all* fish use and is far too important to be left to the so-called consensus of the few, namely those who believe that the world is pretzel-shaped.

Chapter 8
Fish Feel Pain? The Scientists' 10-Point Response

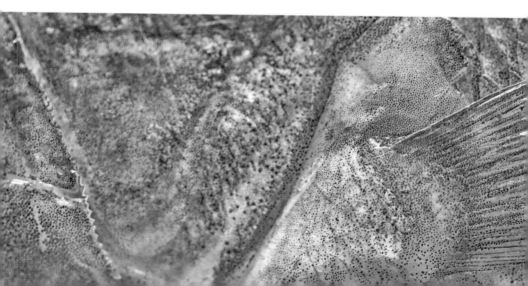

FISH FEEL PAIN!

Reasons to be skeptical about sentience and pain in fishes and aquatic invertebrates

Benjamin K Diggles[a], Robert Arlinghaus[b], Howard I Browman[c], Steven J Cooke[d], Robin L Cooper[e], Ian G Cowx[f], Charles D Derby[g], Stuart W Derbyshire[h], Paul JB Hart[i], Brian Jones[j], Alexander O Kasumyan[k], Brian Key[l], Julian G Pepperell[m], D Christopher Rogers[n], James D Rose[o], Alex Schwab[p], Anne B Skiftesvik[c], Don Stevens[q], Jeffrey D Shields[r], Craig Watson[s]

[a]DigsFish Services, Banksia Beach QLD, Australia; [b]Department of Fish Biology, Fisheries and Aquaculture, Leibniz Institute of Freshwater Ecology and Inland Fisheries & Humboldt, Universität zu Berlin, Berlin, Germany; [c]Institute of Marine Research, Austevoll Research Station, Norway; [d]Department of Biology, Carleton University, Ottawa, ON, Canada; [e]Department of Biology, University of Kentucky, Lexington, KY, USA; [f]Hull International Fisheries Institute, University of Hull, UK; [g]Neuroscience Institute and Department of Biology, Georgia State University, Atlanta, Georgia, USA; [h]Department of Psychology, National University of Singapore, Singapore; [i]Department of Psychology and Vision Science, University of Leicester, Leicester, UK; [j]School of Veterinary Science, Murdoch University, Perth, WA, Australia; [k]Department of Ichthyology, Faculty of Biology, Lomonosov Moscow State University, Moscow, Russia; [l]School of Biomedical Sciences, University of Queensland, Australia; [m]Pepperell Research and Consulting, Noosaville, QLD, Australia; [n]Kansas Biological Survey, and the Biodiversity Institute, The University of Kansas, Lawrence, KS, USA; [o]Emeritus (Retired) Department of Zoology and Physiology, University of Wyoming, Laramie, WY, USA; [p]Britannia Heights, Nelson, New Zealand; [q]Biomed Sci, Atlantic Veterinary College, University of PEI, Charlottetown, Canada; [r]The College of William & Mary, Virginia Institute of Marine Science, Gloucester Point, Virginia, USA; [s]Emeritus (Retired) Tropical Aquaculture Laboratory, University of Florida, Gainesville, USA

Article information:

Title: Reasons to be skeptical about sentience and pain in fishes and aquatic invertebrates

Journal: Reviews in Fisheries Science & Aquaculture

Publisher: Taylor & Francis Online

Publication date: 4 October 2023

Copyright: Open access

To cite this article:

Benjamin K. Diggles, Robert Arlinghaus, Howard I. Browman, Steven J. Cooke, Robin L. Cooper, Ian G. Cowx, Charles D. Derby, Stuart W. Derbyshire, Paul J.B. Hart, Brian Jones, Alexander O. Kasumyan, Brian Key, Julian G. Pepperell, D. Christopher Rogers, James D. Rose, Alex Schwab, Anne B. Skiftesvik, Don Stevens, Jeffrey D. Shields & Craig Watson (2023) Reasons to Be Skeptical about Sentience and Pain in Fishes and Aquatic Invertebrates, Reviews in Fisheries Science & Aquaculture, DOI: 10.1080/23308249.2023.2257802

or

Diggles et al (2023). Reasons to be skeptical about sentience and pain in fishes and aquatic invertebrates. Reviews in Fisheries Science & Aquaculture. https://doi.org/10.1080/23308249.2023.2257802

CONTACT Brian Key @ brian.key@uq.edu.au @ School of Biomedical Sciences, University of Queensland, Australia.
© Supplemental data for this article can be accessed online at https://doi.org/10.1080/23308249.2023.2257802.

Abstract

The welfare of fishes and aquatic invertebrates is important, and several jurisdictions have included these taxa under welfare regulation in recent years. Regulation of welfare requires use of scientifically validated welfare criteria. This is why applying Mertonian skepticism toward claims for sentience and pain in fishes and aquatic invertebrates is scientifically sound and prudent, particularly when those claims are used to justify legislation regulating the welfare of these taxa. Enacting welfare legislation for these taxa without strong scientific evidence is a societal and political choice that risks creating scientific and interpretational problems as well as major policy challenges, including the potential to generate significant unintended consequences. In contrast, a more rigorous science-based approach to the welfare of aquatic organisms that is based on verified, validated and measurable endpoints is more likely to result in "win-win" scenarios that minimize the risk of unintended negative impacts for all stakeholders, including fish and aquatic invertebrates. The authors identify as supporters of animal welfare, and emphasize that this issue is not about choosing between welfare and no welfare for fish and aquatic invertebrates, but rather to ensure that important decisions about their welfare are based on scientifically robust evidence. These ten reasons are delivered in the spirit of organized skepticism to orient legislators, decision makers and the scientific community, and alert them to the need to maintain a high scientific evidential bar for any operational welfare indicators used for aquatic animals, particularly those mandated by legislation. Moving forward, maintaining the highest scientific standards is vitally important, in order to protect not only aquatic animal welfare, but also global food security and the welfare of humans.

Keywords

Animal ethics; fisheries; aquaculture; sentience; suffering; welfare; verification; policy

Introduction

Some countries, including the United Kingdom (UK), have recently passed legislation recognizing fish and selected aquatic invertebrates (decapod crustaceans and cephalopods) as "sentient beings" requiring protection under animal welfare regulations (Birch et al. 2021; Crump et al. 2022; Moylan 2022). Given their important ecological and food production roles (Golden et al. 2021; FAO 2022; Tigchelaar et al. 2022), more interest in fish and aquatic invertebrate welfare is welcome. Nevertheless, extending welfare legislation to more and more animal groups, often following intense lobbying by activist non-government organizations (NGOs) is problematic, as the evidence used to support this move includes advocacy-based reviews (not meta-analyses) which provide contentious interpretations of a limited body of scientific evidence that has significant technical and interpretational flaws (reviewed in Rose et al. 2014; Key 2015; Browman et al. 2019; Diggles 2019; Mason and Lavery 2022; Hart 2023). In fact, the reviews of Birch et al. (2021) and Crump et al. (2022) are being interpreted so broadly and so uncritically in certain policy-making circles that legislators risk following a path ending with all animals including insects and even plankton, eventually being included in welfare legislation (Gibbons et al. 2022; Crump et al. 2022, 2023). The ramifications of these developments on food production, and on how humans interact with animals, would be profound and far reaching (Browman et al. 2019).

Historically, most animal welfare legislation worldwide was based on a suffering-centered framework focussed on the needs of individual terrestrial vertebrates in food production or laboratory settings, specifically those that are sentient and, therefore, may suffer during experimentation, husbandry, and slaughter (Arlinghaus et al. 2009; Browman et al. 2019). Bringing aquatic species under the same suffering-centered legislation frameworks might seem logical at first, but it risks a rush to legislate that may sideline many well-established physiological, pathological, nutritional, and health-related performance indicators that are currently used to operationally define welfare for these animal groups (i.e., criteria used by the functional, pragmatic approach to aquatic animal welfare, see Arlinghaus et al. 2009; Browman et al. 2019). Given the uncertainty surrounding the quality and interpretation of the available science (see Michel 2019; Mason and Lavery 2022 for detailed accounts), applying suffering-centered criteria to the welfare of fishes and aquatic invertebrates under existing legislative frameworks presents major policy challenges and has the potential to generate significant unintended consequences for both animal and human welfare (Browman et al. 2019). The risk of unintended consequences to food security is already being recognized in some countries, leading to societal reluctance and political unwillingness to enact the revised legislation (e.g., https://www.bbc.com/news/science-environment-65691321; https://www.politico.eu/article/uk-conservatives-abandon-plan-protect-animal-welfare/)

Presented here are ten scientifically sound and prudent reasons to maintain skepticism (in the Mertonian sense) when the topics of sentience and pain in fish and aquatic invertebrates are used to justify inclusion of these organisms in legislation governing their use in the wild, food production or research (Table 1).

To be absolutely clear, the authors identify as supporters of animal welfare: this article is not about choosing between no welfare, and welfare for the animal groups involved. Including more animal taxa under welfare legislation is a societal choice that can be undertaken without firm scientific evidence using the precautionary approach. To do so while avoiding significant unforeseen consequences to food security and human welfare, however, requires reliable, scientifically proven operational welfare indicators that uphold the fundamental principles and standards required of the scientific method (Table 2). These ten reasons are presented not as an exhaustive review of the relevant literature, but as a guide to scientists and policy makers to raise awareness of the limitations of the available science about sentience and pain in aquatic animals (Sutherland et al. 2013), and to encourage application of higher scientific standards of evidence to any new operational welfare indicators for aquatic taxa which may be prescribed in welfare legislation. Each of the ten reasons is anchored in case studies or relevant literature and based on real-world examples.

The discussions presented here are underpinned by the philosophy that science advances by conjectures and refutations (Popper 1963). If it is hypothesized that fish, crabs or octopuses are sentient, then predictions must be derived from that hypothesis and tested by experimentation. Only when the various predictions have passed the appropriate tests can it be said that the null hypothesis has been disproven, although complete certainty may never be achieved. The nature of consciousness as experienced by humans is such that testing predictions in animals which are evolutionarily far removed from humans is particularly challenging (Hart 2023). As will be explored in what follows, there is still a long way to go before a satisfactory consensus can be reached regarding sentience of fishes and aquatic invertebrates.

Ten reasons to be skeptical
i. Changing definitions

As will become apparent, it is important to establish a common baseline for several definitions. The meaning of "pain" in English has been relatively consistent throughout history, referring to physical suffering or discomfort caused by illness or injury. A formal definition of the word "pain" was established to describe the human emotional experience that is often, but not always, associated with trauma or injury (https://www.iasp-pain.org/resources/terminology/). The most recent definition endorsed by the International Association for the Study of Pain (IASP) in 2020 describes pain as *"An unpleasant sensory and emotional experience associated with, or resembling that associated with, actual or potential tissue damage"*. This wording was revised from the previous definition because it could have excluded those unable to verbally articulate their pain such as infants, elderly people, and non-human animals (Raja et al. 2020; IASP 2020). Pain is, therefore, the ability of an individual to experience an emotional response to tissue damage or trauma that is detected *via* a process called nociception. Here nociception is defined as the non-conscious processing of noxious stimuli by the peripheral and central nervous system (Tracey 2017).

Pain and nociception are different phenomena (Rose et al. 2014; Tracey 2017). Pain is but one (of many) potential responses to nociception, and merely observing an animal's detection of and response to a stimulus cannot automatically be interpreted as pain. For this reason, the IASP specifically notes that *"Pain cannot be inferred solely from activity in sensory neurons"* (IASP 2020). Nevertheless, some researchers working in the field of fish and invertebrate pain and welfare have developed their own criteria for defining and assessing animal pain based on a range of neurological, behavioral and motivational criteria that they believe is consistent with *"the idea of pain"* (Sneddon et al. 2014; Walters 2018a; Elwood 2021; Sneddon and Roques 2023). These unconventional definitions "shift the goalposts" considerably, as many of the criteria (e.g., avoidance behavior) cannot discriminate between nociception and pain, with some behaviors (e.g., those arising from exposure to chemicals in water) not even requiring nociception. These inconsistencies undermine confidence that interpretation of animal behavioral reactions "consistent with pain" is analogous to how the word pain is defined, used, and understood by humans in accordance with the IASP definition. This has been particularly problematic in the fish and crustacean welfare literature in which any behavior in response to noxious stimuli is usually interpreted as "consistent with pain" (e.g., Elwood 2021; Sneddon and Roques 2023), with few, if any, of the several other alternative explanations being considered (see Rose et al. 2014; Key 2015; Diggles 2019; Browman et al. 2019 for detailed accounts).

The gold standard for pain is verbal reporting. For humans unable to communicate (such as infants, or patients with dementia), pain is assessed by examining motor responses and brain activity in response to noxious stimuli known to produce pain in healthy adult subjects.

Similar comparative approaches are often adopted when attempting to infer pain in non-human animals. Use of the word "pain", however, becomes progressively less defensible as taxa further and further away in evolutionary and morphological terms from humans are considered, because it becomes progressively less plausible that there is an equivalent psychological experience across those taxa. Philosophically speaking, humans struggle to "read the minds" of evolutionarily distant animals (Mameli and Bortolotti 2006), as it is very easy to overinterpret their reactions from a human perspective (Hart 2023). This is particularly so for invertebrates because the neuroanatomy and physiology of insects, crustaceans and molluscs are vastly different to those of mammals (Eisemann et al. 1984; Zullo and Hochner 2011; Walters 2018b; Key et al. 2021), but the same issues also apply to the > 33,000 species of fish which have evolved a vast array of highly specialized sensory systems (Nelson et al. 2016). The closest common ancestor between the invertebrates and early chordates is thought to have lived more than 550 million years ago (Walters 2018b), and while fish do possess a central brain for processing information, in the various aforementioned invertebrate groups many behaviors occur *via* peripheral processing in multiple neural centers without requiring involvement of a central brain (e.g., Ayali 2009; Derby and Thiel 2014; Smarandache-Wellmann 2016; Kuuspalu et al. 2022; Chang and Hale 2023). For these reasons, it is reasonable to ask how analogous (if at all) their experiences to noxious stimuli are to the human experience, and thus how relevant phylogenetically retrospective use of the word pain becomes in these groups (Derbyshire 2016; Walters 2018a, 2018b, 2022; Diggles 2019; Hart 2023). Furthermore, an evolutionary prerequisite for pain in these taxa is not necessarily required, as nociception (or other sensory pathways through which an organism can impart nocifensive reactions or learn avoidance responses to noxious or life-threatening stimuli), is sufficient to guarantee fitness and be selected upon. Importantly, since current legislation regulating the welfare of animals is based on pain (or related concepts such as suffering) and not nociception, caution must be applied when the word pain, or the phrase "consistent with pain", is transposed with nociception without unequivocal justification (see Browman et al. 2019).

The discipline of ethology has developed an extensive set of terms intended to avoid anthropomorphism and observer bias when describing animal behavior (e.g., Bolgan et al. 2015). For example, when it is uncertain that a particular animal group is sentient or can feel pain and psychologically suffer, the term nociception is used (Mason and Lavery 2022). Using the word pain in association with behaviors for which nociception may not necessarily be required (such as avoidance behavior), in species where sentience is impossible to unequivocally prove, is therefore not scientifically appropriate. This is because transposing "pain" for nociception is a "red herring" (Mason and Lavery 2022), in that it presupposes sentience. It is particularly inappropriate to make such a transposition when interpreting experiments that are used to assess whether a particular animal group is sentient. For these reasons, all due skepticism should be applied to research on the welfare of fishes and invertebrates that uses terms that unnecessarily stray from ethological neutrality and replaces them with loaded words or terms used in human psychology. All of this is an extravagant form of anthropomorphism (Rose et al. 2014) that invites false equivalence between the experiences (if any) of those animals and that of human pain (Derbyshire 2016; Hart 2023).

Sentience is the "what it is like" experience of sensory stimuli (e.g., feeling pain and pleasure) (Nagel 1974). Fish and aquatic invertebrates such as molluscs, decapod crustaceans, as well as their arthropod relatives the insects, have a variety of sensory organs that allow them to respond to various stimuli (food odors, predator cues, pheromones etc.) in their environment and to learn from them to maintain fitness (survival and reproduction) under natural selection. Nevertheless, their responses to potentially life threatening stimuli do not necessarily demonstrate awareness or prove pain, consciousness or sentience. For example, the reaction of certain fish species to olfactory detection of chemical alarm cues is usually manifested as flight and hiding behavior, accompanied by physiological stress (Rehnberg and Schreck 1987; Rehnberg et al. 1987; Wisenden 2015). Despite all these behaviors, there is no reason to believe that fish reacting to chemical alarm cues experience pain or suffering. Even so, some political scientists and philosophers have recently introduced "eight sentience criteria" to assess the presence of pain and sentience based on "confidence levels" which supposedly consider both the amount of evidence for a claim, and the reliability and quality of the scientific work behind the evidence (e.g., Birch et al. 2021; Crump et al. 2022, 2023). There are many problems with this approach, including difficulties associated with deciding whether, for example, a rubbing behavior (e.g., Sneddon 2003; Dickerson 2006) is self-protective when that behavior might be reflexive, like the stridulation by spiny lobsters after harmless capture (Bouwma and Herrnkind 2009). Loose interpretation of behaviors, combined with the high number of criteria used, essentially means that there is a very high chance that no organism will fail to meet their threshold for "some evidence of sentience" simply through their ability to sense the surrounding environment (Walters 2022). This type of sentience framework is tautological and, as such, violates the scientific method, because it cannot be refuted. These examples demonstrate that in the socio-political world of decision making about which animal groups should be protected under animal welfare regulations, *"The devil is in the definitions"*.

ii. Ignoring or dismissing conflicting or contradictory evidence

One characteristic of the literature claiming to support the existence of sentience and pain in fishes and invertebrates has been a certain tendency to ignore studies that either do not support this conclusion, or which fail to replicate certain experimental outcomes (Hart 2023). For example, Birch et al. (2021) concluded that cephalopods and decapod crustaceans are sentient, feel pain, and suffer. Yet in the case of crustaceans those authors did not mention the evidence-based reviews of Browman et al. (2019) and Diggles (2019) that concluded the opposing view, and, more importantly, ignored the many scientific problems and unresolved issues identified therein that greatly weaken the evidence used to support their conclusions. Rather, two newly published reviews that concluded crustaceans are sentient and feel pain (Conte et al. 2021; Passantino et al. 2021) were quickly embraced by activist NGOs during the "Crustacean Compassion" campaign and were cited by Birch et al. (2021), despite the fact that they also selectively ignored studies that do not support their conclusions, and overinterpreted and misrepresented other literature (including using strawman arguments). Purportedly authoritative and comprehensive reviews that "cherry-pick" literature to support their narrative fall well short of the scientific rigor that is needed to underpin policy decisions that have widespread implications. It is equally important to consider where such papers were published, because some journals have different editorial criteria, processes, and publication standards (Beall 2017; Grudniewicz et al. 2019; Crosetto 2021; Oviedo-Garcia 2021).

Similarly, a recent review paper on anesthesia in decapod crustaceans (Valente 2022) also excluded or misrepresented some of the literature, and used biased and emotive language, including an abstract which began with the statement *"Decapod crustaceans are sentient beings, not only responding to noxious stimuli but also being capable of feeling pain, discomfort, and distress"*. To avoid a downward spiral in scientific objectivity, rigorous and transparent systematic meta-analyses should be employed using structured methods that are routine in biomedicine and the biological sciences (Mulrow 1994; Dobrow et al. 2004; Aromataris and Pearson 2014; Clements et al. 2022). These methods have not yet been widely applied in the fish and invertebrate welfare literature (see Dawkins 2006; Cooke 2016).

Similar problems of excluding contradictory evidence also occur in the fish welfare literature. The widespread anecdote that hooking fish during angling is painful and hook removal requires analgesics stems from a relatively small body of scientific work (Sneddon 2003; Sneddon et al. 2003; Mettam et al. 2011). This anecdote is contradicted, however, by the studies of Eckroth et al. (2014), Pullen et al. (2017) and Hlina et al. (2021) which found no significant differences between control and treatment groups of Atlantic cod (*Gadus morhua*), northern pike (*Esox lucius*) or bluegill sunfish (*Lepomis macrochirus*) (respectively) exposed to fishing hooks and/or chemicals injected into the mouth. Hlina et al. (2021) found that bluegill that were hooked and then unhooked under controlled laboratory conditions did not significantly differ in their behavior compared with control fish that were not hooked at all. Only those bluegill that were hooked and retained the hook differed in behavior from controls, exhibiting increased use of a shelter. Similarly, hooked Atlantic cod exhibited head shaking but otherwise no other measurable response (Eckroth et al. 2014), which was mirrored by the result of Pullen et al. (2017) who found that northern pike with retained lures did not show behavioral or physiological reactions that differed from controls. The studies of Eckroth et al. (2014), Pullen et al. (2017) and Hlina et al. (2021) all indicate that these fish are resilient and do not significantly alter their behavior in response to acute tissue damage associated with hooking injury to the mouth. This is a consistent result found by disparate research groups that is largely ignored in the suffering-centered fish welfare literature (e.g. Sneddon and Roques 2023 which does not cite Eckroth et al. (2014), despite the two papers sharing at least one co-author).

Perhaps these contradictory studies are being ignored in certain literature because they indicate the behavioral criteria being advocated for defining "pain" behavior in fishes (as well as crustaceans and other invertebrates) are context specific, and, therefore, likely to be inconsistent and unreliable, especially under real world conditions outside of the laboratory. For example, some individual carp (*Cyprinus carpio*), which experience stress from catching by hook and line can remember this event for some time. This occurs through a process of insight learning, however this ability varies with different strains of carp as well as other factors (Beukema 1970a), while the memory of the experience itself is only transitory (Czapla et al. 2023). Northern pike also have the ability to learn hook avoidance, but only if they are caught on an artificial lure; such a memory is not formed if they are caught using live bait (Beukema 1970b). Moreover, the same protective insight learning occurs in many other fish species after a single exposure to a combination of a previously indifferent stimulus with a chemical alarm cue, an odor signal that cannot cause pain (Wisenden 2015). In yet other fish species (including some salmonids), recognition of predator chemical cues is innate and this instinctive response requires no learning process, while other species (including fathead minnows *Pimephales promelas*) show varying degrees of innate bias (Wisenden 2015).

The relative indifference to trauma to the mouth of fishes during angling (at least compared to a normal human exposed to a similar situation) has been recognized for centuries (Cholmondeley-Pennell 1870) and appears remarkable at first glance, but may be explained by neurological evidence. Studies have shown that the trigeminal nerves of teleost fishes have a very low (4–5%) percentage of unmyelinated "C type" or "C fibre" nociceptors (Sneddon 2002) which are responsible for transmitting nociceptive information that may result in sustained and diffuse burning or dull pain in humans (Rose et al. 2014). Furthermore, in elasmobranchs it appears that C type fibers may be absent altogether (Snow et al. 1993; Smith and Lewin 2009), even though sharks show similar behaviors to teleost fish (e.g., fleeing reaction) when hooked by anglers. The latter is also a widely recognized observation known for centuries, which again calls into question referring to pain states from behavior alone.

In contrast to fishes, around 80% of the cutaneous nerve fiber population in normal mammalian nerves are C type fibers. In some types of congenital insensitivity to pain in humans, however, C type fibers are reduced to around 20% of the axon population, with the remainder being the A-delta type (Rosemberg et al. 1994; Guo et al. 2004). Rose et al. (2014) discussed the functional significance of this extremely small percentage of C fibers in teleosts, which is around five times less than in humans with complete insensitivity to pain and up to 20 times less than a normal human. They concluded that *"It appears most logical to assume that in teleosts, at least those species that have been studied, A-delta afferents serve to signal potentially injurious events rapidly, thereby triggering escape and avoidance responses, but that the paucity of C fibers that mediate slow, agonizing, second pain* and pathological pain states (in organisms capable of consciousness) is not a functional domain of nociception in fishes" (Rose et al. 2014). The conclusions of Rose et al. (2014) remain valid, and provide context to the available evidence from Eckroth et al. (2014), Pullen et al. (2017) and Hlina et al. (2021) as well as that of the saline injected fishes from Sneddon (2003), Sneddon et al. (2003) and Mettam et al. (2011), all of which also exhibited no "pain behaviors".

Rose et al. (2014) noted that *"Embedding a fish hook is comparable with the mechanical tissue damage caused by embedding a needle of similar size, but without the saline injection"*. The above cited studies which demonstrate a minimal and often even no impact of fish hooks on fishes represent a consistent body of evidence that debunks the theory of fish pain from hooking. This evidence also calls into question any definition which accepts pain as "an animal's response to stimuli that would be painful for a human" (e.g., see Fiorito et al. 2015). The latter is clearly a "double standard", especially considering that invertebrate taxa like crustaceans and cephalopods naturally undertake behaviors that are completely alien to humans, such as autotomy (shedding of limbs), autophagy (eating one's own body parts), auto-mutilation during essential processes such as reproduction (e.g., Budelmann 1998), as well as regrowth of lost limbs (Murayama et al. 1994; Mariappan et al. 2000).

To summarize, quality evidence-based decision-making weighs all scientific evidence for a particular hypothesis in a rigorous, critical, balanced, transparent, and systematic manner. It does not overinterpret context-specific findings (Rose et al. 2014; Hart 2023), exclude valid studies that report contradictory outcomes (Eckroth et al. 2014; Pullen et al. 2017; Hlina et al. 2021), or fail to consider alternative well-established neurological or behavioral mechanisms which are often not controlled for, yet could also explain the results obtained (Diggles et al. 2017; Walters 2018a; Diggles 2019; Kuuspalu et al. 2022; Chang and Hale 2023). Nor does it recognize as valid studies that have fundamental experimental design flaws such that their conclusions are not warranted (Cooke 2016; Smaldino and McElreath 2016; Key et al. 2017).

iii. Lack of replicable empirical evidence

High quality scientific research produces results that are replicable and independently verifiable by multiple research groups. The "replication crisis" (Ioannidis 2005; Camerer et al. 2018; Clark et al. 2020; Yang et al. 2020, 2023) is also relevant to the fish and invertebrate welfare literature (Rose et al. 2014; Browman et al. 2019; Diggles 2019; Walters 2022; Hart 2023).

In the crustacean welfare literature, a typical example of replication failure was demonstrated by Puri and Faulkes (2010). These researchers failed to replicate an earlier study by Barr et al. (2008) who described grooming and rubbing of prawn antennae exposed to acids (vinegar), bases and an anesthetic as *consistent with the idea of pain*". Indeed, Puri and Faulkes (2010) failed to find any evidence that crayfish or shrimp antennae even had nociceptors that detect acids or bases, suggesting that Barr et al. (2008) had mischaracterized normal grooming (Bauer 1981; Puri and Faulkes 2010), or even other behaviors normally initiated by chemosensors (such as olfaction or gustation, see Johnson and Ache 1978; Derby and Weissburg 2014; Diggles 2019) as evidence of nociception and "pain". It is notable that this lack of verification did not dissuade Birch et al. (2021) from retaining grooming as a welfare criterion for crustaceans, or even acknowledging the ambiguity of the interpretation of Barr et al. (2008). Similar experimental methods and behavioral "pain criteria" have also been applied to cephalopods (e.g., Crook 2021), however the relatively nascent state of research on cephalopod nociception means that the conclusions of Crook et al. await confirmation by other research groups. This is particularly important because all of the same issues and problems with interpretation and replication discussed above for crustaceans (and below for fishes) also apply to cephalopods, especially considering that it is well known that individual cephalopods can react to the same stimulus with quite different responses (Borrelli et al. 2020).

In the case of fishes, Newby and Stevens (2008, 2009) failed to replicate several key results of early fish "pain" research conducted by Sneddon (2003) and Sneddon et al. (2003), something that continues to be ignored by some (e.g., Elwood 2021; Sneddon and Roques 2023). More recently, Rey et al. (2015) claimed that they found evidence for "emotional fever" (stress induced hyperthermia, SIH) in zebrafish (*Danio rerio*) and stated that "...*this finding removes a key argument for lack of consciousness in fishes*". Although these extraordinary claims received considerable media coverage at the time, many technical and interpretational problems meant that the study failed to provide the evidence required to support such claims (Key et al. 2017). The concerns of Key et al. (2017) were later confirmed by Jones et al. (2019), who found no evidence of stress-induced hyperthermia in zebrafish. Importantly, a recent study that used the same tank array as Rey et al. (2015) also failed to replicate stress induced hyperthermia in zebrafish (Vera et al. 2023), while curiously not citing the critical work by Jones et al. (2019). It is notable that the scientific self-correction process that rectified the wildly overinterpreted claims of Rey et al. (2015) took nearly five years. It is relatively easy to make an unfounded claim, but usually far more difficult to scientifically refute it through further experimentation. This demonstrates the asymmetry of this process, especially in the spotlight of the modern media cycle where the initial headlines vastly overshadow the small font used for any eventual correction or retraction.

Rose et al. (2014), Browman et al. (2019) and Diggles (2019) provide many other examples from the fish and invertebrate welfare literature where subsequent studies failed to replicate earlier results obtained through use of unverified and unvalidated welfare indicators. Efforts to draw attention to these instances of overinterpretation are notable because the research community can usually predict which results are unlikely to be replicable (Camerer et al. 2018), because they can identify "*low-powered research coupled with bias selecting for significant results for publication*" (Camerer et al. 2018; see also Clark et al. 2020; Clements et al. 2022; Yang et al. 2023).

In summary, it is vitally important that any new operational welfare indicators used to underpin legislation which defines best practice guidelines for the welfare of fish and invertebrates are valid, robust, measurable, consistent under varying environmental conditions (particularly in the real world outside the laboratory) and independently replicable and verifiable (Table 2).

iv. *Ad hominem* attacks on skeptics

Ad hominem attacks, while common, are a fallacious form of argumentation that should not be part of science. When the evidence for fish and invertebrate pain was found wanting (see Reasons 2 and 3 for selected examples), some of the researchers who conclude that fish are sentient and feel pain attempted to direct attention away from the evidence that contradicted their assertions by discrediting skeptical scientists and labeling them "deniers" (see Diggles and Browman 2018), "creationists" (Sneddon 2013) or even "racists" (Vettese et al. 2020). Demeaning and factually incorrect attacks in pseudo-journals, online, or in the popular media are cleverly potent, because they are often repeated unquestioningly by those unfamiliar with the underlying science, particularly in social media forums, special interest groups and the press. Unfortunately, such activities also lead to completely unacceptable outcomes, including attempts at public shaming by activist groups, and even anonymous threats of violence and intimidation which create a climate of fear amongst skeptics, who understandably wish to avoid such constant attacks. Most concerning, however, is a recent example in an undergraduate textbook (Orth 2023) where several demonstrably false statements were attributed to "skeptics", including that fish were "*incapable of complex cognitive abilities*", and that skeptics "*oppose the need for regulations governing the welfare of fish*". These fallacies are strawman arguments that distort an opposing position into an extreme or weakened version of itself, so that proponents can argue against the newly manufactured position.

Another example is a critique of evidence for fish pain (Key 2016a) in *Animal Sentience*, a publication established in 2015 by the Humane Society of the United States. Responses to Key's invited critique from the readership of that publication included many attacks on the author, rather than his arguments. What is often misunderstood during these asymmetrical, *ad hominem* attacks and the subsequent unfounded claims of "scientific consensus" about fish pain, is that reproducible science eventually triumphs. Science does not advance simply by counting the number of attackers (also colloquially known as a "pile on" or the "bandwagon effect", see Key 2016b) in order to claim "consensus" in an unrepresentative and unreviewed forum (Brown 2016). Instead, scientific advancement depends on the merits of solid, replicable evidence (Abbot et al. 2023).

Attempts to silence scientific debate over the need for reliable empirical evidence for decision making is a form of "cancel culture" which has increasingly pervaded public debate in recent years. An underlying cancel culture theme can be detected in the article by Crump et al. (2022) where they advocate for bans on cephalopod aquaculture and the sale of live decapod crustaceans to private individuals, as a "*low-cost intervention to improve welfare*". Logically, such a position would also result in legal problems for fishers possessing live cephalopods or crustaceans they have caught, and could initiate bans on sale of live crustaceans and cephalopods in the ornamental (pet) aquarium industry and in public aquaria. It is unfortunate to see calls for such bans being repeated in the literature (e.g., see Wuertz et al. 2023). If similar standards were applied to the handling of ornamental fishes, birds or mammals, it would eliminate these companion animal industries overnight, resulting in increased numbers of euthanized, stray or uncared-for animals, while depriving humans of the many health benefits arising from their pets (Beck and Meyers 1996; Brooks et al. 2018) in what could only be described as a "lose-lose" scenario for animals and humans alike. It is important to remember that often the most ardent supporters for welfare of animals are the users of those animals, which for aquatic animals includes recreational anglers (Shephard et al. 2023).

Indeed, banning certain animal uses as advocated by Crump et al. (2022) is a hallmark of the animal rights movement, which is intrinsically opposed to all animal use (Arlinghaus et al. 2009; Arlinghaus and Schwab 2011). Interrogating the ethics of how animals are treated by humans is a philosophical endeavor. In contrast, animal welfare is based on objective scientific evidence which enables principles to be derived that allow animal use for human benefit, whilst recognizing the need for ethical treatment of those animals. This is achieved by providing people and industries with the tools and guidelines to maximize animal welfare (Arlinghaus and Schwab 2011; Stoner 2012; Fiorito et al. 2014, 2015; Diggles 2016). True animal welfare thus provides the potential to develop "win-win" scenarios that benefit both humans and animals, unlike the high potential for generating "lose-lose" scenarios inherent in the application of animal rights ideology. Therefore, to avoid far-reaching "lose-lose" consequences of adopting an animal rights approach to cephalopod aquaculture and the handling of live crustaceans, a more balanced science-based welfare-oriented approach provides advice to industry and consumers on best practice for rearing, handling and dispatch methods for these taxa (as has already been done for fishes, e.g., Cooke et al. 2013; Cook et al. 2015; Diggles 2016). The latter approach certainly has a lower cost to the economy and livelihoods, while supporting the human right to avoid hunger (Golden et al. 2021; FAO 2022; https://www.un.org/sustainabledevelopment/hunger/), all without adding unnecessary burden to handling or processing of fresh seafood.

v. Testing of unfalsifiable hypotheses

As was emphasized at the start of this article, an important cornerstone of the self-corrective aspect of the scientific method is hypothesis falsifiability (Popper 1963). This is an important point, because most of the research conducted on fish and invertebrate sentience, pain and suffering tests hypotheses that are unfalsifiable (Browman and Skiftesvik 2011). For example, Sneddon and Roques (2023) state that *"Responses to 'painful treatment' will differ between species and between individuals"*. Sneddon and Roques (2023) also state:

> General indicators, such as the overall physical condition of the fish, the presence of lesions, demeanor, and body or fin posture, make a contribution to the assessment of pain, but they alone do not determine whether the animal is in pain. Pain is inherently stressful and, as such, physiologic indicators of stress can assist in understanding the extent to which pain affects welfare and homeostasis. More importantly, changes in biological function traits can be used more effectively to determine if an animal is pain-free; if the animals exhibit normal behavior and demeanor, no significant stress responses, are healthy and disease-free, reproduce normally, and grow normally, then there is likely no pain.

These statements suggest that, by these definitions, virtually any behavioral changes in fish could be interpreted as indicating "pain", whether from a specific noxious stimulus or otherwise. This position ignores important differences between nociception, stress, and pain (Stevens et al. 2016, also see Reason 1) and makes the question of pain in fishes a non-falsifiable hypothesis. In other words, Sneddon and Roques (2023) argue that any deviation from what an observer believes is "normal" behavior for that particular individual fish is a result of pain. This approach basically assumes that all "normal" behaviors in each fish species are known and quantifiable, that there are known ranges of "normal" and "painful" behavior for individual fish, and that there are no explanations other than pain for any behavioral deviations from "normal".

The same problems also occur in the crustacean literature where, for example, some behaviors deemed *"consistent with the idea of pain"* (Barr et al. 2008; Elwood 2021) are most likely an overinterpretation or misrepresentation of normal grooming or chemosensory behaviors (Puri and Faulkes 2010; Diggles 2019). Experiments on cephalopods have also revealed a tendency for individual animals to react to the same stimulus with quite different responses, such that standardization of testing protocols is urgently required (Borrelli et al. 2020) if there is to be any chance of consistent application of the scientific method toward validating welfare criteria for these animals.

This is not a trivial issue. If unfalsifiable hypotheses based on unvalidated criteria are used as the basis for drafting welfare legislation encompassing fishes, crustaceans and other invertebrates, any activities that an observer believes deviates from their qualitative idea of "normal" behavior could be interpreted as "painful", and hence could meet legislative criteria for infringement and prosecution. Similarly, if unfalsifiable criteria are used to judge whether organisms are sentient or feeling pain, it is difficult to see how any organism would fail to meet criteria requiring their protection under welfare legislation, unless the criteria were arbitrarily applied.

vi. Arbitrary application of criteria

The criteria advocated by Birch et al. (2021) to ascribe sentience to animals are being applied arbitrarily. For example, Birch et al. (2021) argue, based on their criteria, that all cephalopods and decapod crustaceans should now be considered "sentient beings", yet within the Mollusca they do not extend their analyses to other groups such as bivalves (e.g., scallops, oysters) and gastropods (e.g., abalone, snails). These taxa also react in response to visual, chemical, noxious and environmental cues (e.g., Barnhart et al. 2008; Wesołowska and Wesołowski 2014; Siemann et al. 2015; Hochner and Glanzman 2016; Walters 2018b, 2022) including alleged "avoidance learning" (Selbach et al. 2022), and share similar neuroanatomical networks to the Cephalopoda. Moreover, within the Crustacea, members of the Copepoda have similar physiology and neurological networks to the Decapoda and also react in response to visual, chemical, noxious and environmental cues. Based on the criteria of Birch et al. (2021), copepods, bivalves, and gastropods would appear to satisfy at least three or four of their eight criteria with reasonably high certainty, leading to a potentially erroneous conclusion of "some evidence" or "substantial evidence" of sentience in these groups (Walters 2022).

Perhaps these criteria are being applied arbitrarily because taking their consistent application to its logical conclusion would be extremely problematic. For example, sea lice (*Lepeophtheirus* spp., *Caligus* spp.) are ectoparasitic copepods that cost the global salmon farming industry hundreds of millions of dollars annually to control (Abolofia 2017; Stene et al. 2022). This cost is incurred in large part to satisfy animal welfare concerns over the impact of lice infestation on the welfare of wild and cultured salmon (Macaulay et al. 2022), but without any regard for the impact of the treatments on the welfare of the sea lice (Moccia et al. 2020). Similarly, extending the same sentience analysis to bivalve molluscs could result in bans on the consumption of fresh, live oysters.

The problems with the criteria of Birch et al. (2021) do not stop there. Single cell protozoans exhibit nociception yet they have no cell-based nervous system; changes in behavior during predator avoidance are triggered by changes in bioelectrical activity within cell membranes (Naitoh 1974; Valentine and Van Houten 2022) or by response to chemical gradients (King and Insall 2009). Even bacteria exhibit not only quorum sensing but also learning behavior "*similar to Pavlovian conditioning*" (Hopkin 2008), while slime molds are alleged to exhibit learning and problem-solving behaviors (Bonner 2010; Boussard et al. 2019). Indeed, such behavior is not restricted to microorganisms from the animal kingdom, given that plants make sounds when stressed by dehydration (Khait et al. 2023; Marris 2023).

Essentially, all of this means that if the criteria used by Birch et al. (2021) were universally applied, there is a high chance that few, if any, organisms would fail to meet their threshold for "some evidence of sentience" (Walters 2022). Such an outcome brings the utility and validity of the criteria themselves into serious question; if all organisms are considered "sentient", this severely devalues the feelings-based welfare concept itself into irrelevance, because everything (and therefore nothing) is special all at the same time (Birch 2017). Thus, at some stage there would still be a need for further arbitrary decision making regarding which groups deserve welfare protection, and where or when exemptions must be applied to preserve the human population's health and food supply systems (e.g., in wild capture fisheries, use of insecticides to protect crops or control mosquitoes to combat malaria or other vector-borne diseases). Sentience is a "hard problem" (Gray 2004; Humphrey 2022; Mason and Lavery 2022), and these severe shortcomings of the sentience criteria used by Birch et al. (2021) highlight why their utility is questionable, as is more broadly, application of the feelings-based suffering-centered approach to the welfare of fish and aquatic invertebrates. Instead of arbitrary application of intangible, context-dependent concepts that may selectively serve certain ethical positions (no use of animals at all) at the cost of others (sustainable use of animals), pragmatic functional or nature-based operational welfare indicators with quantifiable endpoints that are scientifically validated, reliable and straightforward to interpret should be used (Arlinghaus et al. 2009; Diggles et al. 2011; Barragán-Méndez et al. 2019; Browman et al. 2019; Table 2).

vii. Dilution/devaluation of the welfare concept

A broad definition of sentience is being extended from terrestrial vertebrates into not only fish, crustaceans, molluscs, and other invertebrates including insects (Crump et al. 2022, 2023; Gibbons et al. 2022), but also plants (Calvo et al. 2017; Chamovitz 2018; Baluška and Mancuso 2021) and cell cultures (Niikawa et al. 2022). Most of this discussion is philosophical in nature, since measuring "pleasure" and "pain" in these groups in any scientifically valid context remains challenging and riddled with inconsistencies, technical problems (e.g., Bennett et al. 2009; Borrelli et al. 2020) and subjective anthropomorphic assumptions (e.g., for fishes see Rose et al. 2014; Mason and Lavery 2022; Hart 2023; and for plants see Brown and Key 2021).

The major problems with assuming all animals (and increasingly plants) are sentient until proven otherwise were identified by Birch (2017) as being *"unscientific or anti-scientific"*, and that such a position would make *"the science of animal sentience more or less irrelevant to the scope of animal protection law: all animals would be assumed sentient unless proven otherwise, and it is hard to see how research could prove otherwise"* (Birch 2017). This proposition by Birch would require proving a negative, almost an impossibility, but properly done research could determine if a species failed to meet sound, testable criteria for sentience if they were to become available.

The argument that bees are sentient is illustrative of some of the problems of adopting this approach. If the potentially unfalsifiable hypothesis that bees are sentient is accepted (Gibbons et al. 2022; Crump et al. 2023), ignoring the fact that their behaviors can be replicated by robots (Adamo 2019), it follows that all insects are potentially sentient and should therefore be theoretically protected under welfare regulation. This situation would lead to widespread noncompliance with welfare legislation when farmers need to protect their crops from predatory insects. A likely (and presumably unintended) consequence of such an event would be severe constraints on crop production imposed by bans on insecticides to protect insects and regulation of tilling the soil because of all of the sentient animals displaced and killed in doing so. Issues would also arise when trying to protect bees. For example, parasitic mites (arachnid arthropods) have been implicated in the worldwide collapse of bee colonies. Can mites be killed using allegedly "painful" procedures if it helps save bees? What if mites turn out to be more sentient than bees? (Reber 2017).

Given that arthropod parasites (including crustaceans) and other microorganisms are common disease agents (or vectors for diseases) of animals and humans, it appears extremely implausible that any laws would ever be passed to protect their welfare. What would be the impact on animal welfare of treatments for arthropod parasites such as bedbugs, mosquitoes, headlice, fleas, skin mites and ticks? If sentience, and therefore, welfare protection were extended to helminth parasites such as nematodes and cestodes, the deliberate poisoning of intestinal worms would raise ethical issues that would need to be ignored by those afflicted. As would the more mundane problems of killing tens of thousands of allegedly sentient beings when mowing the lawn or driving your car along a country road.

Less obvious threats to biodiversity would also arise. For example, the culture of the critically endangered freshwater mussel would raise animal ethics issues because the larval stage of this bivalve mollusc is an obligate parasite on the gills of freshwater fish (Barnhart et al. 2008). Is it still ethical to save the mollusc by deliberately exposing or directly infecting fish? The latter strategy has been widely accepted since the beginning of the twentieth century in an attempt to prevent loss of biodiversity (Buddensiek 1995).

If the intent of sentience proponents is to protect more animal taxa under welfare legislation, there is no need to dilute welfare concepts, and generate "lose-lose" situations for animals and humans while denigrating the scientific method in the process. As already pointed out, these taxa can be included in welfare frameworks under pragmatic functional or nature-based welfare definitions (Arlinghaus et al. 2009; Diggles et al. 2011) which can make welfare a "win-win" scenario for both animals and humans alike. Otherwise, how can welfare for fishes or invertebrates be considered in instances such as endocrine disruption, for example, where it has been demonstrated that entire populations of fishes can collapse (Kidd et al. 2007) when exposed to estrogenic chemical pollutants? Individual animals affected by endocrine disruption do not necessarily "suffer" in a conventional sense (Diggles et al. 2011), so a suffering centered approach to welfare will not suffice.

Poor water quality and damaged habitat can lead to complete loss of fish or invertebrate populations. No fish or invertebrates means no fish or invertebrate welfare. Solving this equation for fish and invertebrates, however, yields the following: water quality and habitat = fish and invertebrate welfare, thus providing a potential "win-win" for both humans and animals. As demonstrated by the biodiversity crisis and endocrine disruption problems, in practice, workable animal welfare frameworks need to be able to align with ecological reality in the natural world. To achieve this in the Anthropocene, animal welfare needs to be reframed as ecological welfare with an integrated approach to general ethical principles that also encompass ecological, environmental and societal issues (Fox 2006) such as biodiversity protection, human and animal health and food security (Golden et al. 2021; FAO 2022; Macaulay et al. 2022; Troell et al. 2023; Allen et al. 2023). If food webs collapse, so will human society, at which point nothing is served by demanding a welfare status for an individual animal, be it a fish, crab, snail, cephalopod or insect. The "One Health" approach to sustainable food system design advocated by the World Health Organization appears to be an appropriate starting point signaling the way forward in this regard (Stentiford et al. 2020).

viii. High risk of unintended consequences

There are always risks of unintended consequences when governments regulate, and welfare legislation is no exception. A notable example followed the inclusion of decapod crustaceans in welfare regulations in Victoria, Australia in June 2019 (Supplement 1). In this case, Royal Society for Prevention of Cruelty to Animals (RSPCA) field officers threatened fines and prosecution of a restaurateur who was housing live mud crabs (*Scylla serrata*) in a display aquarium, because the crabs had their claws tied to their bodies. This was considered a breach of the *Prevention of Cruelty to Animals Act 1986* as the RSPCA inspector considered that the crabs "*needed to be allowed to move their arms freely.*" The tying method used had been implemented by the live mud crab industry for many decades based not only on occupational health and safety concerns (to prevent people from getting their fingers and hands crushed by crabs), but also the fact that free claws greatly increase claw autotomy rates and allows the mud crabs to injure, kill and eat other mud crabs held in the same display tanks due to their naturally cannibalistic nature.

This case study highlights how unvalidated, anthropomorphic, feelings-based welfare criteria applied to new animal groups under suffering-centered animal welfare legislation frameworks can result in retrograde "lose-lose" welfare outcomes for the animals involved, as well as injury and prosecution of innocent people. Crump et al. (2022) appeal for *"more research into appropriate stocking densities, environmental conditions, and methods to prevent aggression and injury"* in captive decapod crustaceans. This is laudable, however they overlook the fact that decades of practical experience in "quality management" have already developed that information as well as many other reliable physiological, pathological, nutritional, and health related welfare indicators for holding live decapod crustaceans (e.g., Paterson and Spanoghe 1997; Davidson and Hosking 2004; Shields et al. 2006). This large body of existing evidence underpins animal health, survival and profitability in many fisheries and aquaculture industries; hence it should not be ignored in a rush for "new" feelings-based welfare criteria for decapod crustaceans.

Eyestalk ablation to boost larval production from penaeid shrimp broodstock has also emerged as a welfare issue in recent years. Diarte-Plata et al. (2012) suggested that ablation was "painful" based on tail flicking and leg or antennal rubbing as welfare indicators. Neither tail flicking nor rubbing are validated or reliable pain indicators in crustaceans, however, as shown by Puri and Faulkes (2010) in the case of rubbing and Weineck et al. (2018) who demonstrated that tail flicking is a reflex that also occurs in transected shrimp abdomens separated from the head. Nevertheless, eyestalk ablation affects several other easily quantifiable functional welfare metrics such as broodstock survival and larval quality; hence alternatives to ablation are used by the shrimp aquaculture industry when they are available (Magaña-Gallegos et al. 2018).

Recent studies of the shrimp *Penaeus vannamei* have found alternatives to eyestalk ablation that can result in equal or better quality and quantity of larvae for that species (Zacarias et al. 2019, 2021). This result is not universal, however, and to date no alternatives to ablation have been found for other important cultured shrimp species such as *Penaeus monodon* (see Uddin and Rahman 2015). Demands by certain interest groups to ban eyestalk ablation in all shrimp farming would result in the use of ten to twenty times more *P. monodon* broodstock to meet industry needs for post larvae. This would immediately have the unintended consequence of requiring many more *P. monodon* broodstock, another "lose-lose" situation as it conflicts with one of the basic 3Rs welfare principles of reduction of numbers of animals used. Such a move would also increase fishing pressure on wild stocks, while the lack of reliable larval supply would threaten entire aquaculture industries in countries where *P. vannamei* is not available, threatening livelihoods and regional and/ or global food security.

Similar issues have arisen in the regulation of commercial fishing in the European Union (EU). For example, attempts to increase gear selectivity and reduce the environmental impact of beam trawling for flatfishes in the Netherlands resulted in development of electric pulse trawling (Kraan et al. 2020; Penca 2022). Instead of using tickler chains which damage sensitive benthic habitats, the electric pulse trawl method uses a suspended hydrofoil generating pulsed electric fields (up to 60 volts, 30–45 Hz) within the net area to generate galvanotaxis of benthic fishes and increase the potential for their capture in the trawl net without interacting directly with the sediment (Kraan et al. 2020). Intensive scientific scrutiny found electric pulse trawling was more selective for flatfish and brown shrimp (Verschueren et al. 2019; Poos et al. 2020), reduced bycatch and benthic disturbance, and also reduced fishing time and fuel consumption/CO_2 emissions, resulting in higher sustainability scores compared to fishing with traditional beam trawls (Kraan et al. 2020; Penca 2022). Paradoxically, despite this, animal welfare concerns were amongst the reasons provided when electric pulse trawling was banned by the EU in 2019 (Court of Justice of the European Union 2021; Penca 2022).

The precautionary approach (see Reason 9 below) as well as political and socio-economic considerations relating to encroachment of the pulse trawl fleet into areas historically fished by low impact netting methods, were some of the main drivers of the decision (Penca 2022). Nevertheless, intense activity from activist NGOs also highlighted perceived welfare concerns for flatfish as well as larger Atlantic cod which may experience spinal damage during capture when exposed to the electric pulses (Kraan et al. 2020). Atlantic cod typically have a low survival rate if they are caught in trawls, so the issue of spinal damage was considered irrelevant from a fisheries sustainability perspective (Kraan et al. 2020), assuming of course that the fishery itself is appropriately managed. Nevertheless, the court decision was upheld upon appeal, supported by false and misleading claims by activist NGOs who *"compromised the basic principles of ethical conduct in scientific research"* (Kraan et al. 2020). Some aspects of this decision were considered likely to stifle technological innovation in the fisheries sector which would otherwise improve conservation, reduce overfishing, and improve fisheries sustainability (Kraan et al. 2020; Penca 2022).

Another "lose-lose" example is the banning of catch-and-release angling in some countries, such as Germany and Switzerland. In both these countries, voluntary catch and release of legal-sized fish by anglers was considered by some members of the community to be unethical and cruel, as the caught fish could be eaten rather than released (Arlinghaus et al. 2012). After intensive lobbying by animal rights groups, mandatory catch-and-retain regulations were implemented, which result in more fish being killed by anglers (Arlinghaus et al. 2012). Increased fishing mortality can result in potential reductions of both recruitment and sustainability in fisheries which generate substantial socio-economic activity and provide a range of human health benefits (Arlinghaus et al. 2007). A mandatory catch-and-retain policy based on animal rights ideology therefore paradoxically results in reduction of not only the welfare of populations of fish (due to the removal of fish that would otherwise continue to live), but also a reduction in human welfare as well.

All of the above examples provide glimpses into the range of unintended consequences and potential "lose-lose" scenarios that could be expected if a suffering-centered approach, fueled by animal rights advocacy, is allowed to influence welfare legislation for fishes and aquatic invertebrates, and that legislation is subsequently enacted in the absence of a firm scientific basis.

ix. Dangers of the precautionary approach

Various definitions of the precautionary approach exist, using wording such as *"Where there are threats of serious or irreversible damage, lack of full scientific certainty shall not be used as a reason for postponing cost-effective measures to prevent that damage"* (Principle 15 of the 1992 Rio Declaration). The decision to invoke the precautionary approach is generally associated with risk assessment of both the seriousness and magnitude of the supposed threat whilst considering any inherent uncertainties (Krebs 2011). As shown in Reasons 6–8, however, the precautionary approach is often invoked by activist NGOs in order to more aggressively weaponize their campaigns for change.

What needs to be understood by policy makers is when the precautionary approach is used to prioritize the welfare needs of individual (allegedly sentient) species which are ecologically lower in the food chain (such as smaller fishes and aquatic invertebrates which are natural prey items for larger predatory fishes or aquatic invertebrates), several predator/prey conundrums arise. In aquaculture, one issue is the use of live feeds. The use of crustacean zooplankton (i.e., *Artemia*, copepods) as live feeds is an essential prerequisite for the health, welfare and survival of virtually all larval fish, decapod crustaceans and cephalopods. A pragmatic science and nature-based approach to fish and invertebrate welfare supports live feeding on the grounds that live feeds form the basis of natural food webs, and hence provide the best nutrition, health and welfare outcomes for larval fish and invertebrates. Provided that live feed cultures are biosecure, microbially clean and disease free, they are accepted as necessary and critical for replicating natural feeding processes that underpin the entire global aquaculture industry (Støttrup and McEvoy 2003).

In stark contrast, Crump et al. (2022) use what are essentially animal rights definitions to identify live feed as a welfare problem for the prey animals. This shows how the precautionary use of a suffering-centered welfare framework does not mirror reality in any predator/prey situation that supports natural trophic pathways (i.e., in natural environments, wild fisheries or aquaculture nurseries), because it only recognizes the needs of individual animals ("dots"), but it is unable to "join the dots" together *via* trophic pathways into a coherent working ecosystem (Fox 2006). "Joining the dots" also involves humans, who are part of the same natural world where we are all inextricably linked to the other dots through our health, welfare and food security needs (Allen et al. 2023). Regulating live feeding for fisheries or aquaculture under a suffering-centered welfare framework, usually leads to circular arguments about benefits for predators versus prey. This argument is eventually self-limiting, however, because animal rights definitions eventually conclude that to resolve the conundrum the solution is to humanely kill all predators (Fox 2006; Bramble 2021), or genetically engineer them to become vegetarian (Fox 2006). Of course, in the real world, both so called "solutions" would undermine essential trophic and evolutionary processes and risk global ecological collapse (Allen et al. 2023).

Similar animal rights-based arguments are also being made against fish feeds based on insect meal,[1] despite this natural protein source resulting in better welfare for carnivorous fish than feeding them terrestrial plant-based diets (Alfiko et al. 2022). Recent calls

---

[1] https://thefishsite.com/articles/anti-insect-ingredient-farmed-salmon-standard-tops-new-welfare-benchmark

by animal rights groups and others to ban farming of octopus and other cephalopods, allegedly based on welfare grounds (Schnell et al. 2022), also invoke the precautionary approach, to the arguable detriment of overexploitation of wild cephalopod populations.[2] Utilizing the precautionary approach to advocate for banning aquaculture production of predatory fish and invertebrates based on perceived welfare issues with live feeds under suffering-centered welfare frameworks would result in immediate global food insecurity (Golden et al. 2021; Tigchelaar et al. 2022; Troell et al. 2023). On the other hand, playing devil's advocate with the very same precautionary approach could lead to an argument that to reduce these clear risks to global food security and avoid other presumably unintended consequences, all aquatic taxa should be removed from all animal welfare protection. Of course, this extreme example would likely be unacceptable to society, as well as potentially counterproductive to the health and welfare of aquatic animals. But as pointed out by Arlinghaus and Schwab (2011), the precautionary approach or "benefit of the doubt", can neutralize everything, including common sense.

x. The need for organized skepticism and critical thinking

The scientific method is based upon observations, hypotheses to test and in turn understand the observations, data collection, and analysis of repeatable empirical evidence for or against the hypotheses. The method then requires rigorous skepticism regarding the empirical evidence (Huxley 1866; Popper 1963), and self-correction through defining specific criteria for falsifying the hypotheses (Trevors 2010; Abbot et al. 2023). As such, science is best seen as organized skepticism: *"a journey, over time, toward*

---

[2] https://www.theguardian.com/environment/2023/jun/25/a-symbol-of-what-humans-shouldnt-be-doing-the-new-world-of-octopus-farming

contingent understanding guided by experimental tests and skeptical questioning" (May 2011).

The examples in the previous nine points demonstrate that the scientific method is being regularly misused, misinterpreted, or ignored in the fish and invertebrate sentience and pain literature. Moreover, some of this literature (and other associated publication activities) contains evidence of organized activism. The strategies being used by some groups to advocate for legislative changes ignore several of the Mertonian principles of science (Abbot et al. 2023). This pathway usually leads to pseudoscience, which if left unchallenged, promotes Lysenkoism-like activity (Gordin 2012; Kolchinsky et al. 2017) leading to a slippery slope which unravels the fundamentals of all sciences, regardless of discipline.

One of the most important defenses against pseudoscience is critical thinking. A good way to explain critical thinking is *via* a worked example. Imagine Ms. B. Leave and Ms. A. Gainst are engaged in a discussion about the existence of the tooth fairy. Ms. B. Leave starts by affirming the existence of the tooth fairy with evidence that her teeth placed on the bedside table at night are replaced with a gold coin the next morning. This didn't happen only once, but many times, and was also verified by other friends who experienced similar events. Wanting direct proof, Ms. A. Gainst proposed to stay awake overnight and witness the tooth fairy for herself, however, Ms. B. Leave exclaimed that the tooth fairy will never come if there is a chance it will be seen. Not surprisingly, Ms. A. Gainst remained skeptical of this non-falsifiable hypothesis. She suggested perhaps there was a more parsimonious explanation, that being Ms B. Leave's parents were the entities responsible for the tooth/coin swap.

The above scenario reveals how four important questions can help guide critical thinking: first, what is the proposed hypothetical mechanism underpinning the phenomenon? i.e., the tooth fairy replaces the tooth with a gold coin. Second, does the evidence support the mechanism? – neither the gold coin nor hearsay is evidence for the tooth fairy. Third, is the mechanism biologically or physically plausible? – flying fairies carrying gold coins are inconsistent with known scientific principles; and fourth, are there other more parsimonious alternative explanations for the phenomenon? e.g., the tooth fairy is instead a real person. These four questions demonstrate the basis of organized skepticism.

Application of organized skepticism and critical thinking to the various scientific issues highlighted here is vitally important if policy makers are to interpret the limited data available (Sutherland et al. 2013) and develop sound policies which avoid the potential "lose-lose" situations and unintended consequences which could otherwise arise (Figure 1). In this age of internet misinformation, "alternative facts" and Artificial Intelligence (AI) written research which fabricates its own references (Davinack 2023), it is also important to maintain and cultivate in our schools and higher education systems a scientific culture of healthy, rigorous, organized skepticism, to protect us from pseudoscience so scientists can continue to generate reliable empirical evidence upon which to base important decisions that affect the future of humankind and animal life (Krebs 2011; May 2011; O'Brien et al. 2021; Abbot et al. 2023). Clearly, the current uncertain state of fish and invertebrate welfare science demands a skeptical view (Mason and Lavery 2022; Hart 2023) to ensure that the scientific record generates reliable knowledge to support evidence-based decision making on this issue of global importance.

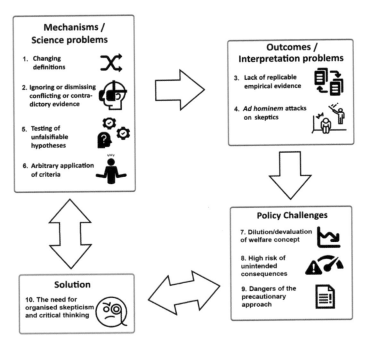

Figure 1. Conceptual diagram of the interactions between the ten reasons to be skeptical about fishes and aquatic invertebrates being sentient and feeling pain. Problems with scientific rigor hinder interpretation of the outcomes, which results in significant policy challenges during development of legislation regulating welfare. To avoid the potential dangers and unintended consequences which can arise from these issues, organized skepticism and critical thinking must be applied to the underlying scientific problems and policy challenges that arise.

Conclusion – a suggested way forward

The ten reasons outlined above emphasize why any new operational welfare indicators used to define best practice guidelines for fish and aquatic invertebrates under suffering-centered welfare frameworks need to be closely scrutinized for their scientific robustness, relevance and applicability. This means they should be held to a scientifically validated evidential standard which makes them measurable, replicable under all conditions (rather than being specific to certain laboratory contexts), reliable and hence equivalent to the well-established physiological, pathological, nutritional, and health-related welfare indicators already used to define current best practice for these animals under the pragmatic animal welfare approach (Table 2). If they are not, it is doubtful that inclusion of fish and invertebrates in animal welfare legislation will generate any meaningful welfare improvement for these animals, at a high risk of "lose-lose" scenarios involving retrograde welfare outcomes and unintended consequences to both humans and aquatic animals alike.

Extending legal protection to fish and aquatic invertebrates is a societal choice (Browman et al. 2019) and politics is the method used in Western democracies to influence that choice (Krebs 2011; Moylan 2022). Nevertheless, science, not politics, is the method best equipped to identify reliable, replicable, and effective operational welfare indicators that can improve welfare outcomes for fish and invertebrates. Effective welfare can be a "win-win" scenario for both aquatic animals and humans alike. Yet without application of organized skepticism, verification and critical thinking to this subject, there is a high risk of invoking a rapid downward spiral in scientific rigor, with potentially significant unintended consequences. This situation would be at odds with the high evidential bar historically applied to the majority of science underpinning development of modern aquaculture and fisheries science and management. With global food security, livelihoods, and the human right to be free from hunger and poverty in play, the stakes are very high. To paraphrase Sir Winston Churchill,

> *Never before in the field of human food security,*
> *has so much been put at risk for so many, based on*
> *so few verifiable facts.*

Table 1. Ten reasons to be skeptical about fishes and invertebrates being sentient and feeling pain

Reason	General principle	Examples
1. Changing definitions	Deviation from accepted definitions of pain. Development of "sentience criteria" based on "confidence levels".	Replacing ethological terms with loaded human psychology terms. Promoting *"the idea of pain"* without considering alternative explanations for observed behavior. Philosophical/political risk analysis approach to defining sentience.
2. Ignoring or dismissing conflicting or contradictory evidence	Selectively ignoring data or studies that are inconsistent with the pain hypothesis.	Accepting evidence that is context specific as being broadly representative of all taxa without consideration of the full range of available evidence. Ignoring existence of "no effect" studies. Ignoring the very low % of C type nociceptors in teleosts (absent in elasmobranchs), circa 5 times less than humans with some forms of congenital insensitivity to pain. Ignoring letters and reviews highlighting significant fundamental flaws in research methodology/interpretation. Role of "Guest Editors" ensuring publication of biased reviews used during activist campaigns.
3. Lack of replicable empirical evidence	Results "consistent with pain and/ or sentience" are not replicable or independently verifiable by multiple research groups.	Grooming and rubbing of prawn antennae exposed to vinegar, bases and anesthetics not repeatable. Rocking and rubbing behavior in rainbow trout not repeatable. Claims of emotional fever in zebrafish as supporting consciousness in fishes not repeatable and shown to be false.

Consequences and outcomes

"Shifts the goalposts" to pain definitions which no longer discriminate between nociception and pain.

Lowers "evidential bar" to include behaviors that do not even require nociception.

Encourages anthropomorphism which invites false equivalence between the experience of animals and that of human pain.

Bias and overinterpretation of context specific data with few/no alternative explanations presented or considered.

Asymmetrically ignoring studies that report contradictory outcomes.

Use of overly emotive language.

Invoking a double standard by accept pain as "animals response to stimuli that would be painful for humans" while ignoring behaviors alien to humans like autotomy, autophagy, auto-mutilation and limb regrowth.

Erosion of normal scientific standards of peer review and publication ethics.

Paucity of rigorous, unbiased and transparent systematic reviews of the relevant literature that include critical appraisal of the evidence base.

Mischaracterization of possible grooming and/or chemosensory behavior in crustaceans as evidence of "pain".

Mischaracterization of possible experimental artifacts (e.g., recovery from anesthesia in fishes) as evidence of "pain".

Overinterpretation of behavioral studies that cannot be replicated by independent research groups.

Potential for new feelings-based operational welfare indicators in fish and crustaceans to be inconsistent, unreliable and/or unverifiable outside the laboratory.

Reason	General principle	Examples
4. *Ad hominem* attacks on skeptics	Attempts to discredit skeptical scientists who highlight flaws in evidence base.	Attacks in pseudo journals or the popular media.
		Labeling skeptical scientists as "deniers", "creationists" or "racists".
		Promoting the "bandwagon effect", to claim a manufactured "consensus" in biased, unrepresentative unreviewed forums or online.
5. Testing of unfalsifiable hypotheses	Testing hypotheses that are unfalsifiable negates the fundamental self-corrective aspect of the scientific method.	Utilizing pain definitions that encompass any "non-normal" behavior or any behavior deemed *"consistent with the idea of pain"*.
		Claims that pain will be expressed differently not only between species but even between individual animals of same species.
		Ignoring important differences between stress, nociception and pain.
6. Arbitrary application of criteria	Selective application of sentience criteria.	Application of sentience criteria to decapod crustaceans but not copepods.
		Application of sentience criteria to cephalopod molluscs but not bivalves or gastropods.
		Apparent unwillingness to apply the same sentience criteria to protozoans, and microorganisms.
7. Dilution/ devaluation of the welfare concept	Welfare protection being extended from terrestrial vertebrates and fishes into crustaceans, insects and other invertebrates, even plants.	If all organisms are considered "sentient" based on alleged pain perception, this severely devalues the feelings based welfare concept itself, because everything (and therefore nothing) is special all at the same time.
		Inability to classify "non-suffering" issues which threaten biodiversity (e.g., endocrine disruption) as a welfare concern.

Attempting to silence scientific debate upon the need for reliable empirical evidence for decision making.

Erroneous claims of "scientific consensus".

Asymmetrical, *ad hominem* public attacks on those seeking replicable evidence.

"Cancel culture" reminiscent of animal rights activism generating "lose-lose" scenarios instead of a balanced "win-win" approach to solving welfare issues in fish and invertebrates.

Switching burden of proof.

Promoting pain definitions which encompass any behavioral changes (whether from a specific noxious stimulus or not).

Using unfalsifiable criteria to underpin legislation means any abnormal behavior could be interpreted as "painful", resulting in infringement and prosecution (with no foreseeable way to prove ones innocence).

Using unfalsifiable criteria to determine whether organisms are sentient will mean all organisms could meet criteria for sentience.

Highly permissive criteria mean that all animals (and possibly even plants) meet threshold for alleged sentience at some level.

Arbitrary application of criteria will be required to maintain relevance and meaning of welfare concept.

These outcomes bring the validity of the criteria themselves into serious question.

If criteria are accepted, further arbitrary decision making will be required to exempt certain activities to preserve the human population's food supply.

A need for widespread noncompliance with welfare legislation to avoid "lose-lose" scenarios for animal and human welfare, and to ensure global food security.

Workable animal welfare frameworks need to be able to align with ecological reality in the natural world.

Need to reframe pragmatic animal welfare principles within an integrated "One health" approach that encompasses welfare as a "win-win" linked with ecological sustainability and global food security.

Reason	General principle	Examples
8. High risk of unintended consequences	Dangers of application of unvalidated welfare criteria to new animal groups under suffering-centered animal welfare legislation frameworks.	Ignoring proven functional welfare indicators in favor of "new" unvalidated feelings based operational welfare indicators. Threatening fines/prosecution of people housing live crabs because crabs had claws tied to body. Demanding bans on eyestalk ablation in broodstock penaeid shrimp. Banning electric pulse trawling due in part because of damage to some larger fish.
9. Dangers of the precautionary approach	Invoking the precautionary approach to act before validated operational welfare criteria have been established.	Misleading claims of equivalence between aquatic species and terrestrial bird and mammalian farm animals. Need for the suffering-centered welfare approach to consider individual sentient animals. Application of suffering-centered welfare frameworks to predator/prey situations results in inability to feed predatory aquatic taxa.
10. The need for organized skepticism and critical thinking	Scientists must understand the limitations of the scientific method and must speak up when the scientific method is being misapplied or ignored.	Misapplication of the scientific method in the context of fish and invertebrate sentience, pain and welfare. Internet misinformation, "alternative facts" and AI written research which fabricates its own references. Blurring/inflation of the science boundary.

Consequences and outcomes

"Lose-lose" scenarios, including retrograde welfare outcomes for both animals and humans.

Banning ablation conflicts with 3R's by requiring use of 10-20 times more broodstock shrimp to achieve same levels of larval production.

Endangering larval supply of entire aquaculture industries would threaten regional/global food security.

Prosecution of innocent people.

Stifling of innovation which could reduce environmental impacts and improve sustainability.

Potential bans on live feeding of larval aquatic animals would impact welfare of fed animals and could shut down entire aquaculture industries overnight, threatening regional/global food security.

Unnecessary bans on farming new taxa (e.g., cephalopods) increase risk of overexploitation of wild populations.

Stifling of innovation which could reduce environmental impacts and improve sustainability.

Pseudoscience, if left unchallenged, promotes Lysenkoism-like activity which harms society.

High risk of "lose-lose" scenarios involving retrograde welfare outcomes and unintended consequences.

Need for organized skepticism and critical thinking to ensure limitations of the scientific method are not exceeded and avoid a rapid downward spiral in scientific rigor.

Robust science is needed to generate reliable empirical data for evidence-based decision making on important topics that affect food security and livelihoods.

Table 2. Summary of replicability, accuracy and reliability for a range of operational welfare indicators for aquatic organisms (excludes molecular, environmental and nutritional parameters).

Taxa	Welfare indicator category	Welfare indicator type	Parameters measured
Finfish	Disease status	External lesions/ deformities	Injury/infection
		Internal/microscopic lesions	Injury/infection
		Parasitic infection	Infection
		Diagnostic pathogen testing	Infection
		Toxicology testing	Contamination
	Performance	Condition factor	Nutritional status
		Food conversion rate	Nutritional efficiency
		Specific growth rate	Weight/size
		Survival rate	Survival
	Physiology	Blood glucose	Activity
		Blood lactate/pH	Activity
		Cortisol (blood/excreted)	Stress
		Haematocrit/cell counts	Immune status
		Heart rate	Many factors
		Immunoglobulins, antibodies, peptides	Immune status
	Behavior	Acoustic activity	Alertness
		Electric activity	Alertness
		Feeding	Appetite
		General activity	Many factors
		Moribund/lethargic	Death
		Opercular beat/gill ventilation rate	Many factors
		Pain	Unknown
		Reproduction	Fecundity/fertilisation
		Righting/tail grab reflex	Alertness, exhaustion
		Rubbing	Many factors
		Rocking	Unknown
		Swimming activity	Many factors
		Vestibulo-ocular reflex	Alertness, exhaustion

✓ = yes, x = no, ± = variable or context specific, ? = unknown/unreliable

Replicable	Verified accurate	Reliability as operational welfare indicator
✓	✓	High
✓	✓	High
±	±	Low/Moderate (depends on parasite)
✓	✓	Low-high (depends on pathogen)
✓	✓	Moderate
✓	✓	High
✓	✓	High
✓	✓	High
✓	✓	High
✓	✓	High
✓	✓	High
✓	✓	High
✓	✓	High
±	±	Moderate
✓	✓	High
±	±	Moderate
±	±	Moderate
✓	✓	High
±	±	Moderate
✓	✓	High
±	±	Moderate
x	x	?
✓	✓	High
✓	✓	High
±	±	Low
x	x	?
±	±	Moderate
✓	✓	High

Taxa	Welfare indicator category	Welfare indicator type	Parameters measured
Crustaceans	Disease status	External lesions/deformities	Shell disease/injury
		Internal/microscopic lesions	Injury/infection
		Parasitic infection	Infection
		Diagnostic pathogen testing	Infection
		Toxicology testing	Contamination
	Performance	Autotomy	Injury/stress
		Food conversion rate	Nutritional efficiency
		Moulting	Growth
		Specific growth rate	Weight/size
		Survival rate	Survival
	Physiology	Crustacean hyperglycemic hormone (CHH)	Hormonal status
		Differential hemocyte count	Infection, immune status
		Haemolymph colour/clotting	Infection
		Haemolymph glucose	Activity, molt status
		Haemolymph lactate	Activity, molt status
		Haemolymph protein	Molt status
		Heart rate	Many factors
		Heat shock protein	Many factors
	Behavior	Autotomy	Many factors
		Feeding	Appetite
		General activity	Many factors
		Grooming	Many factors
		Moribund/lethargic	Death
		Pain	Unknown
		Reproduction	Fecundity/fertilisation
		Righting reflex	Vitality
		Tail flipping	Many factors
		Ventilation rate	Many factors

Replicable	Verified accurate	Reliability as operational welfare indicator
x	±	Low/Moderate (may indicate predator activity)
±	±	Low/Moderate (depends on pathogen)
±	±	Low/Moderate (depends on parasite)
✓	±	Low to high (depends on pathogen)
✓	✓	Moderate
±	±	Low/Moderate (context dependent)
✓	±	Moderate (diet dependent)
✓	✓	High
±	✓	Moderate (variable in captivity)
✓	✓	High
±	±	Moderate (varies throughout molt cycle)
±	±	Low/Moderate (highly variable)
±	±	Low/Moderate (context dependent)
±	±	Low/Moderate (context dependent)
✓	±	Moderate (context dependent)
✓	✓	Moderate (varies throughout molt cycle)
±	±	Moderate (context dependent)
±	±	Low/Moderate (context dependent)
±	±	Low (context dependent)
±	✓	Low/Moderate (varies throughout molt cycle)
±	±	Moderate
x	x	Low
✓	✓	High
x	x	?
✓	✓	High
✓	✓	High
±	±	Low (context dependent)
±	±	Low (context dependent)

Taxa	Welfare indicator category	Welfare indicator type	Parameters measured
Molluscs	Disease status	External lesions	Injury/infection
		Internal/microscopic lesions	Infection
		Parasitic infection	Infection
		Diagnostic pathogen testing	Infection
		Toxicology testing	Contamination
	Performance	Food conversion rate	Nutritional efficiency
		Specific growth rate	Weight/size
		Survival rate	Survival
	Physiology	Differential hemocyte count	Infection, immune status
		Heart rate	Many factors.
		Haemocyanin	Activity, air exposure
		Haemolymph colour/ osmolality	Infection, stress?
		Haemolymph glucose	Activity
		Haemolymph lactate	Activity?
		Haemolymph pH	Activity, air exposure
		Lysozyme	Activity?
	Behavior	Autophagy/ Automutilation	Damaging/eating oneself (in cephalopods)
		Colour	Colour change (not applicable to bivalves)
		Grooming	Many factors
		Inking	Expulsion of ink (not applicable to bivalves)
		Pain	Unknown
		Reproduction	Fecundity/fertilisation
		Ventilation rate	Many factors
		Righting reflex	Vitality (cephalopods, gastropods)

Replicable	Verified accurate	Reliability as operational welfare indicator
±	±	Low/Moderate (may indicate predator activity or senescence)
±	±	Low/Moderate (depends on pathogen)
±	±	Low/Moderate (depends on parasite)
±	±	Low to high (depends on pathogen)
✓	✓	Moderate
±	±	Low/Moderate (diet dependent, difficult to ascertain in filter feeders)
±	±	Moderate (depends on species)
✓	✓	High
±	±	Low/Moderate (highly variable)
±	±	Low (highly variable)
±	?	Low (context dependent)
±	?	Low (context dependent)
±	?	Low (context dependent)
±	?	Low (context dependent)
±	±	Low (context dependent)
±	?	Low (context dependent)
±	?	Low (context dependent)
±	?	Low (context dependent and highly variable)
±	±	Low (context dependent)
±	±	Low (context dependent)
x	x	?
✓	✓	High
±	±	Low (context dependent)
✓	✓	High

Acknowledgements

We thank Rohan Pethiyagoda, Zen Faulkes and three anonymous reviewers for helpful comments and critiques of earlier drafts of the manuscript.

Author contributions

Conceptualization: SJ Cooke, R Arlinghaus.

Writing, figures – HI Browman, SJ Cooke, BK Diggles, PJB Hart, JB Jones, B Key, DC Rogers, A Schwab, JD Shields, AB Skiftesvik, ED Stevens.

Review and editing: R Arlinghaus, HI Browman, SJ Cooke, RL Cooper, IG Cowx, CD Derby, SW Derbyshire, BK Diggles, PJB Hart, JB Jones, AO Kasumyan, B Key, JG Pepperell, DC Rogers, JD Rose, JD Shields, A Schwab, AB Skiftesvik, ED Stevens, C Watson.

Disclosure statement

This paper was written without direct funding by a broad global cross section of research scientists who work in a diverse range of fields involving research on fish and aquatic invertebrates as well as on applied aspects of fisheries and aquaculture. All of us declare an interest in upholding high scientific standards on all topics, particularly those relating to sustainable food production, fisheries and conservation of biodiversity. Many of the authors (14/20) fish for food and/or recreation, while all of the authors eat seafood, otherwise there are no relevant competing interests to declare. The opinions and positions taken in this article are those of the authors, and do not necessarily reflect those of their employers.

ORCID

Brian Key ⓘ http://orcid.org/0000-0002-1150-3848

References

Abbot D, Bikfalvi A, Bleske-Rechek AL, Bodmer W, Boghossian P, Carvalho CM, Ciccolini J, Coyne JA, Gauss J, Gill PMW, et al. 2023. In defense of merit in science. Controversial Ideas 3(1):1. doi:10.35995/jci03010001.

Abolofia J, Asche F, Wilen JE. 2017. The cost of lice: quantifying the impacts of parasitic sea lice on farmed salmon. Mar Res Econ. 32(3):329–349. doi:10.1086/691981.

Adamo SA. 2019. Is it pain if it does not hurt? On the unlikelihood of insect pain. Can Entomol. 151(6):685–695. doi:10.4039/tce.2019.49.

Allen BL, Bobier C, Dawson S, Fleming PJS, Hampton J, Jachowski D, Kerley GIH, Linnell JDC, Marnewick K, Minnie L, et al. 2023. Why humans kill animals and why we cannot avoid it. Sci Total Environ. 896:165283. doi:10.1016/j.scitotenv.2023.165283.

Alfiko Y, Xie D, Astuti RT, Wong J, Wang L. 2022. Insects as a feed ingredient for fish culture: status and trends. Aquacult Fish. 7(2):166–178. doi:10.1016/j.aaf.2021.10.004. Arlinghaus R, Schwab A. 2011. Five ethical challenges to recreational fishing: What they are and what do they mean? Amer Fish Soc Symp. 75:219–234.

Arlinghaus R, Schwab A. 2011. Five ethical challenges to recreational fishing: What they are and what do they mean? Amer Fish Soc Symp. 75:219–234.

Arlinghaus R, Cooke SJ, Schwab A, Cowx IG. 2009. Contrasting pragmatic and suffering-centred approaches to fish welfare in recreational fishing. J Fish Biol.Biol. 75(10):2448–2463. doi:10.1111/j.1095-8649.2009.02466.x.

Arlinghaus R, Schwab A, Riepe C, Teel T. 2012. A primer on anti-angling philosophy and its relevance for recreational fisheries in urbanized societies. Fisheries. 37(4):153–164. doi:10.1080/03632415.2012.666472.

Aromataris E, Pearson A. 2014. The systematic review: an overview. Am J Nurs. 114(3):53–58. doi:10.1097/01.NAJ.0000444496.24228.2c.

Ayali A. 2009. The role of the arthropod stomatogastric nervous system in moulting behaviour and ecdysis. J Exp Biol. 212(Pt 4):453–459. doi:10.1242/jeb.023879.

Baluška F, Mancuso S. 2021. Individuality, self and sociality of vascular plants. Philos Trans R Soc Lond B Biol Sci. 376(1821):20190760. doi:10.1098/rstb.2019.0760.

Barnhart MC, Haag WR, Roston WN. 2008. Adaptations to host infection and larval parasitism in Unionoida. J Nth Amer Benthol Soc. 27(2):370–394. doi:10.1899/07-093.1.

Barr S, Laming PR, Dick JT, Elwood RW. 2008. Nociception or pain in a decapod crustacean? Anim Behav. 75(3):745– 751. doi:10.1016/j.anbehav.2007.07.004.

Barragán-Méndez C, Sobrino I, Marín-Rincón A, Fernández-Boo S, Costas B, Mancera JM, Ruiz-Jarabo I. 2019. Acute-stress biomarkers in three Octopodidae species after bottom trawling. Front Physiol. 10:784. doi:10.3389/fphys.2019.00784.

Bauer RT. 1981. Grooming behavior and morphology in the decapod Crustacea. J Crust Biol. 1(2):153–173. doi:10.2307/1548154.

Beall J. 2017. What I learned from predatory publishers. Biochem Med. 27(2):273–278. doi:10.11613/BM.2017.029.

Beck AM, Meyers NM. 1996. Health enhancement and companion animal ownership. Annu Rev Public Health. 17(1):247–257. doi:10.1146/annurev.pu.17.050196.001335.

Bennett CM, Miller M, Wolford G. 2009. Neural correlates of interspecies perspective taking in the post-mortem Atlantic salmon: an argument for multiple comparisons correction. Neuroimage. 47:S125. doi:10.1016/S1053-8119(09)71202-9.

Beukema JJ. 1970a. Angling experiments with carp (*Cyprinus carpio* L.) II. Decreasing catchability through one-trial learning. Neth J Zool. 20(1):81–92. doi:10.1163/002829670X00088.

Beukema JJ. 1970b. Acquired hook-avoidance in the pike *Esox lucius* L. fished with artificial and natural baits. J Fish Biol. 2:155–160. doi:10.1111/j.1095-8649.1970.tb03268.x.

Birch J. 2017. Animal sentience and the precautionary principle. Anim Sent. 2(16):1200. doi:10.51291/2377-7478.1200.

Birch J, Burn C, Schnell A, Browning H, Crump H. 2021. Review of the evidence of sentience in cephalopod molluscs and decapod crustaceans. Report by the London School of Economics and Political Science. https://www.lse.ac.uk/business/consulting/reports/review-of-the-evidence-of-sentiences-in-cephalopod-molluscs-and-decapod-crustaceans.

Bolgan M, O'Brien J, Gammell M. 2015. The behavioural repertoire of Arctic charr (*Salvelinus alpinus* (L.)) in captivity: a case study for testing ethogram completeness and reducing observer effects. Ecol Freshw Fish. 25(2):318–328. doi:10.1111/eff.12212.

Bonner JT. 2010. Brainless behavior: a myxomycete chooses a balanced diet. Proc Natl Acad Sci USA. 107(12):5267– 5268. doi:10.1073/pnas.1000861107.

Borrelli L, Chiandetti C, Fiorito G. 2020. A standardized battery of tests to measure *Octopus vulgaris*' behavioural performance. Invert Neurosci. 20(1):4. doi:10.1007/s10158-020-0237-7.

Boussard A, Delescluse J, Perez-Escudero A, Dussutour A. 2019. Memory inception and preservation in slime moulds: the quest for a common mechanism. Philos Trans R Soc Lond B Biol Sci. 374(1774):20180368. doi:10.1098/rstb.2018.0368.

Bouwma PE, Herrnkind WF. 2009. Sound production in Caribbean spiny lobster *Panulirus argus* and its role in escape during predatory attack by *Octopus briareus*. NZ J Mar FW Res. 43(1):3–13. doi:10.1080/00288330909509977.

Bramble B. 2021. Painlessly killing predators. J Applied Philosophy 38(2):217–225. doi:10.1080/00288330909509977.

Brooks HL, Rushton K, Lovell K, Bee P, Walker L, Grant L, Rogers A. 2018. The power of support from companion animals for people living with mental health problems: a systematic review and narrative synthesis of the evidence. BMC Psychiatry. 18(1):31. doi:10.1186/s12888-018-1613-2.

Browman HI, Skiftesvik AB. 2011. Welfare in aquatic organisms – is there some faith-based HARKing going on here? Dis Aquat Organ. 94(3):255–257. doi:10.3354/dao02366.

Browman HI, Cooke SJ, Cowx IG, Derbyshire SWG, Kasumyan A, Key B, Rose JD, Schwab A, Skiftesvik AB, Stevens ED, et al. 2019. Welfare of aquatic animals: where things are, where they are going, and what it means for research, aquaculture, recreational angling, and commercial fishing. ICES J Mar Sci. 76(1):82–92. doi:10.1093/icesjms/fsy067.

Brown C. 2016. Fish pain: an inconvenient truth. Anim Sent. 2016:058.

Brown D, Key B. 2021. Plant sentience, semantics, and the emergentist dilemma. J Consci Stud. 28:155–183.

Buddensiek V. 1995. Culture of juvenile freshwater pearl mussels *Margaritifera margaritifera* L. in cages: a contribution to conservation programmes and the knowledge of habitat requirements. Biol Conserv. 74(1):33–40. doi:10.1016/0006-3207(95)00012-S.

Budelmann BU. 1998. Autophagy in *Octopus*. S Afr J Mar Sci. 20(1):101–108. doi:10.2989/025776198784126502.

Calvo P, Sahi VP, Trewavas A. 2017. Are plants sentient? Plant Cell Environ. 40(11):2858–2869. doi:10.1111/pce.13065.

Camerer CF, Dreber A, Holzmeister F, Ho T-H, Huber J, Johannesson M, Kirchler M, Nave G, Nosek BA, Pfeiffer T, et al. 2018. Evaluating the replicability of social science experiments in *Nature* and *Science* between 2010 and 2015. Nat Hum Behav. 2(9):637–644. doi:10.1038/s41562-018-0399-z.

Chamovitz DA. 2018. Plants are intelligent; Now what? Nat Plants. 4(9):622–623. doi:10.1038/s41477-018-0237-3.

Chang W, Hale ME. 2023. Mechanosensory signal transmission in the arms and the nerve ring, an interarm connective, of *Octopus bimaculoides*. iScience. 26(5):106722. doi:10.1016/j.isci.2023.106722.

Cholmondeley-Pennell H. 1870. Can fish feel pain? The question considered analogically and physiologically with a few words on the ethics of angling. London: Fredrick Warne and Co.

Clark TD, Raby GD, Roche DG, Binning SA, Speers-Roesch B, Jutfelt F, Sundin J. 2020. Ocean acidification does not impair the behaviour of coral reef fishes. Nature. 577(7790):370–375. doi:10.1038/s41586-019-1903-y.

Clements JC, Sundin J, Clark TD, Jutfelt F. 2022. Meta-analysis reveals an extreme "decline effect" in the impacts of ocean acidification on fish behavior. PLoS Biol. 20(2):e3001511. doi:10.1371/journal.pbio.3001511.

Conte F, Voslarova E, Vecerek V, Elwood RW, Coluccio P, Pugliese M, Passantino A. 2021. Humane slaughter of edible decapod crustaceans. Animals. 11(4):1089. doi:10.3390/ani11041089.

Cook KV, Lennox RJ, Hinch SG, Cooke SJ. 2015. Fish out of water: How much air is too much? Fisheries. 40(9):452–461. doi:10.1080/03632415.2015.1074570.

Cooke SJ. 2016. Spinning our wheels and deepening the divide: call for an evidence-based approach to the fish pain debate. Anim Sent. 2016:42.

Cooke SJ, Donaldson MR, O'connor CM, Raby GD, Arlinghaus R, Danylchuk AJ, Hanson KC, Hinch SG, Clark TD, Patterson DA, et al. 2013. The physiological consequences of catch-and-release angling: perspectives on experimental design, interpretation, extrapolation and relevance to stakeholders. Fish Manag Ecol. 20(2–3):268–287. doi:10.1111/j.1365-2400.2012.00867.x.

Court of Justice of the European Union. 2021. Press Release No 59/21. Case C-733/19, Kingdom of the Netherlands v Council of the European Union and European Parliament. https://curia.europa.eu/jcms/upload/docs/application/pdf/2021-04/cp210059en.pdf.

Crook RJ. 2021. Behavioral and neurophysiological evidence suggests affective pain experience in octopus. iScience. 24(3):102229. doi:10.1016/j.isci.2021.102229.

Crosetto P. 2021. Is MDPI a predatory publisher? [accessed 2023 Jan 1]. https://paolocrosetto.wordpress.com/2021/04/12/is-mdpi-a-predatory-publisher/.

Crump A, Browning H, Schnell AK, Burn C, Birch J. 2022. Invertebrate sentience and sustainable seafood. Nat Food. 3(11):884–886. doi:10.1038/s43016-022-00632-6.

Crump A, Gibbons M, Barrett M, Birch J, Chittka L. 2023. Is it time for insect researchers to consider their subjects' welfare? PLoS Biol. 21(6):e3002138. doi:10.1371/journal.pbio.3002138.

Czapla P, Wallerius ML, Monk CT, Cooke SJ, Arlinghaus R. 2023. Re-examining one-trial learning in common carp (*Cyprinus carpio*) through private and social cues: no evidence for hook avoidance lasting more than seven months. Fish Res. 259:106573. doi:10.1016/j.fishres.2022.106573.

Davidson GW, Hosking WW. 2004. Development of a method for alleviating leg loss during post-harvest handling of rock lobsters. Fisheries Research and Development Corporation Project No. 2000/251. https://www.frdc. com.au/sites/default/files/products/2000-251-DLD.PDF.

Davinack AA. 2023. Can ChatGPT be leveraged for taxonomic investigations? Potential and limitations of a new technology. Zootaxa. 5270(2):347–350. doi:10.11646/zoo-taxa.5270.2.12.

Dawkins MS. 2006. A user's guide to animal welfare science. Trends Ecol Evol. 21(2):77–82. doi:10.1016/j.tree.2005.10.017.

Derby CD, Weissburg MJ. 2014. The chemical senses and chemosensory ecology of crustaceans. In: Derby C, Thiel M, editors. Nervous Systems and Control of Behavior, Vol. 3 of The Natural History of the Crustacea (editor-in-chief, M. Thiel). New York: Oxford University Press; pp. 263–292.

Derby C, Thiel M. (editors). 2014. Nervous systems and control of behavior Vol. 3 of The Natural History of the Crustacea (editor-in-chief, M. Thiel). Oxford University Press, New York. ISBN 978-0-19-979171-2.

Derbyshire SW. 2016. Fish lack the brains and psychology for pain. Commentary on Key on Fish Pain. Animal Sent. 2016:025.

Diarte-Plata G, Sainz-Hernández JC, Aguiñaga-Cruz JA, Fierro-Coronado JA, Polanco-Torres A, Puente-Palazuelos C. 2012. Eyestalk ablation procedures to minimize pain in the freshwater prawn *Macrobrachium americanum*. Appl Anim Behav Sci. 140(3–4):172–178. doi:10.1016/j. applanim.2012.06.002.

Dickerson HW. 2006. Chapter 4. *Ichthyophthirius multifiliis* and *Cryptocaryon irritans* (Phylum Ciliophora). In: Woo PTK, editor. Fish Diseases and Disorders Volume 1. CAB International, p. 116–153.

Diggles BK. 2016. Development of resources to promote best practice in the humane dispatch of finfish caught by recreational fishers. Fish Manag Ecol. 23(3–4):200–207. doi:10.1111/fme.12127.

Diggles BK. 2019. Review of some scientific issues related to crustacean welfare. ICES J Mar Sci. 76(1):66–81. doi:10.1093/icesjms/fsy058.

Diggles BK, Browman HI. 2018. Denialism and muddying the water or organized skepticism and clarity? THAT is the question. Anim Sent. 21(10):139. doi:10.51291/2377-7478.1349.

Diggles BK, Cooke SJ, Rose JD, Sawynok W. 2011. Ecology and welfare of aquatic animals in wild capture fisheries. Rev Fish Biol Fisheries. 21(4):739–765. doi:10.1007/s11160-011-9206-x.

Diggles BK, Arlinghaus R, Browman HI, Cooke SJ, Cowx IG, Kasumyan AO, Key B, Rose JD, Sawynok W, Schwab A, et al. 2017. Responses of larval zebrafish to low pH immersion assay. Comment on Lopez-Luna et al. J Exp Biol. 220(Pt 17):3191–3192. doi:10.1242/jeb.162834.

Dobrow MJ, Goel V, Upshur RE. 2004. Evidence-based health policy: context and utilisation. Soc Sci Med. 58(1):207–217. doi:10.1016/s0277-9536(03)00166-7.

Eckroth JR, Aas-Hansen Ø, Sneddon LU, Bichão H, Døving KB. 2014. Physiological and behavioural responses to noxious stimuli in the Atlantic Cod (*Gadus morhua*). PLoS One. 9(6):e100150. doi:10.1371/journal.pone.0100150.

Eisemann CH, Jorgensen WK, Merritt DJ, Rice MJ, Cribb BW, Webb PD, Zalucki MP. 1984. Do insects feel pain? A biological view. Experientia. 40(2):164–167. doi:10.1007/BF01963580.

Elwood RW. 2021. Potential pain in fish and decapods: Similar experimental approaches and similar results. Front Vet Sci. 8:631151. doi:10.3389/fvets.2021.631151.

FAO. 2022. The State of World Fisheries and Aquaculture 2022. Towards Blue Transformation. Rome: FAO. doi:10.4060/cc0461en.

Fiorito G, Affuso A, Anderson DB, Basil J, Bonnaud L, Botta G, Cole A, D'Angelo L, De Girolamo P, Dennison N, et al. 2014. Cephalopods in neuroscience: regulations, research and the 3Rs. Invert Neurosci. 14(1):13–36. doi:10.1007/s10158-013-0165-x.

Fiorito G, Affuso A, Basil J, Cole A, de Girolamo P, D'Angelo L, Dickel L, Gestal C, Grasso F, Kuba M, et al. 2015. Guidelines for the care and welfare of cephalopods in research – a consensus based on an initiative by CephRes, FELASA and the Boyd Group. Lab Anim. 49(2 Suppl):1–90. doi:10.1177/0023677215580006.

Fox W. 2006. Human relationships, nature, and the built environment: problems that any general ethics must be able to address. The MIT Press. http://www.warwickfox.com/files/2007_sage_-_probs_for_ge.pdf.

Gibbons M, Crump A, Barrett M, Sarlak S, et al. 2022. Can insects feel pain? A review of the neural and behavioural evidence. Adv Insect Physiol. 63:155–229. doi:10.1016/bs.aiip.2022.10.001.

Golden CD, Koehn JZ, Shepon A, Passarelli S, Free CM, Viana DF, Matthey H, Eurich JG, Gephart JA, Fluet-Chouinard E, et al. 2021. Aquatic foods to nourish nations. Nature. 598(7880):315–320. doi:10.1038/s41586-021-03917-1.

Gordin MD. 2012. How Lysenkoism became pseudoscience: dobzhansky to velikovsky. J Hist Biol. 45(3):443–468. doi:10.1007/s10739-011-9287-3.

Gray J. 2004. Consciousness. creeping up on the hard problem. Oxford: Oxford University Press.

Grudniewicz A, Moher D, Cobey KD, Bryson GL, Cukier S, Allen K, Ardern C, Balcom L, Barros T, Berger M, et al. 2019. Predatory journals: no definition, no defence. Nature Comment. 576(7786):210–212. doi:10.1038/d41586-019-03759-y.

Guo Y-C, Liao K-K, Soong B-W, Tsai C-P, Niu D-M, Lee H-Y, Lin K-P. 2004. Congenital insensitivity to pain with anhydrosis in Taiwan: a morphometric and genetic study. Eur Neurol. 51(4):206–214. doi:10.1159/000078487.

Hart PJB. 2023. Exploring the limits to our understanding of whether fish feel pain. J Fish Biol. 102(6):1272–1280. doi:10.1111/jfb.15386.

Hlina BL, Glassman DM, Chhor AD, Etherington BS, Elvidge CK, Diggles BK, Cooke SJ. 2021. Hook retention but not hooking injury is associated with behavioral differences in Bluegill. Fish Res. 242:106034. doi:10.1016/j.fishres.2021.106034.

Hopkin M. 2008. Bacteria 'can learn'. Nature 2007:360. doi:10.1038/news.2007.360.

Hochner B, Glanzman DL. 2016. Evolution of highly diverse forms of behavior in molluscs. Curr Biol. 26(20):R965–R971. doi:10.1016/j.cub.2016.08.047.

Humphrey N. 2022. Sentience, The Invention of Consciousness. Oxford: Oxford University Press.

Huxley TH. 1866. On the advisableness of improving natural knowledge. Fortnight Rev. 3:626–637. https://www.gutenberg.org/files/2934/2934-h/2934-h.htm.

IASP. 2020. IASP revises its definition of pain for the first time since 1979. https://www.iasp-pain.org/publications/iasp-news/iasp-announces-revised-definition-of-pain/.

Ioannidis JP. 2005. Why most published research findings are false. PLoS Med. 2(8):e124. doi:10.1371/journal.pmed.0020124.

Johnson BR, Ache BW. 1978. Antennular chemosensitivity in the spiny lobster, *Panulirus argus*: Amino acids as feeding stimuli. Mar Behav Physiol. 5(2):145–157. doi:10.1080/10236247809378530.

Jones NA, Mendo T, Broell F, Webster MM. 2019. No experimental evidence of stress-induced hyperthermia in zebrafish (*Danio rerio*). J Exper Biol. 222:jeb192971.

Key B. 2015. Fish do not feel pain and its implications for understanding phenomenal consciousness. Biol Philos. 30(2):149–165. doi:10.1007/s10539-014-9469-4.

Key B. 2016a. Why fish do not feel pain. Anim Sent. 2016:003.

Key B. 2016b. Burden of proof lies with proposer of celestial teapot hypothesis. Response III to Commentary on Key on Fish Pain. Anim Sent. 2016:079.

Key B, Arlinghaus R, Browman HI, Cooke SJ, et al. 2017. Problems with equating thermal preference with "emotional fever" and sentience. Comment on Rey et al. (2015) Fish can show emotional fever: stress-induced hyperthermia in zebrafish. Proc Roy Soc Lond B. 284:20160681.

Key B, Zalucki O, Brown DJ. 2021. Neural design principles for subjective experience: implications for insects. Front Behav Neurosci. 15:658037. doi:10.3389/fnbeh.2021.658037.

Khait I, Lewin-Epstein O, Sharon R, Saban K, Goldstein R, Anikster Y, Zeron Y, Agassy C, Nizan S, Sharabi G, et al. 2023. Sounds emitted by plants under stress are airborne and informative. Cell. 186(7):1328–1336. e10. doi:10.1016/j.cell.2023.03.009.

Kidd KA, Blanchfield PJ, Mills KH, Palace VP, Evans RE, Lazorchak JM, Flick RW. 2007. Collapse of a fish population after exposure to synthetic estrogen. Proc Natl Acad Sci USA. 104(21):8897–8901. doi:10.1073/pnas.0609568104.

King JS, Insall RH. 2009. Chemotaxis: finding the way forward with *Dictyostelium*. Trends Cell Biol. 19(10):523–530. doi:10.1016/j.tcb.2009.07.004.

Kolchinsky EI, Kutschera U, Hossfeld U, Levi GS. 2017. Russia's new Lysenkoism. Curr Biol. 27:R1037–R1059.

Kraan M, Groeneveld R, Pauwelussen A, Haasnoot T, Bush SR. 2020. Science, subsidies and the politics of the pulse trawl ban in the European Union. Mar Pol. 118:103975. doi:10.1016/j.marpol.2020.103975.

Krebs JR. 2011. Risk, uncertainty and regulation. Philos Trans A Math Phys Eng Sci. 369(1956):4842–4852. doi:10.1098/rsta.2011.0174.

Kuuspalu A, Cody S, Hale ME. 2022. Multiple nerve cords connect the arms of octopuses, providing alternative paths for inter-arm signaling. Curr Biol. 32(24):5415–5421.e3. doi:10.1016/j.cub.2022.11.007.

Macaulay G, Barrett LT, Dempster T. 2022. Recognising trade-offs between welfare and environmental outcomes in aquaculture will enable good decisions. Aquacult Environ Interact. 14:219–227. doi:10.3354/aei00439.

Magaña-Gallegos E, Bautista-Bautista M, González-Zuñiga LM, Arevalo M, Cuzon G, Gaxiola G. 2018. Does unilateral eyestalk ablation affect the quality of the larvae of the pink shrimp *Farfantepenaeus brasiliensis* (Letreille, 1817) (Decapoda: Dendrobranchiata: Penaeidae)? J Crust Biol. 38(4):401–406. doi:10.1093/jcbiol/ruy043.

Mameli M, Bortolotti L. 2006. Animal rights, animal minds, and human mindreading. J Med Ethics. 32(2):84–89. doi:10.1136/jme.2005.013086.

Mariappan P, Balasundaram C, Schmitz B. 2000. Decapod crustacean chelipeds: an overview. J Biosci. 25(3):301–313. doi:10.1007/BF02703939.

Marris E. 2023. Stressed plants 'cry' – and some animals probably hear them. Nature. 616(7956):229–229. doi:10.1038/d41586-023-00890-9.

Mason GJ, Lavery JM. 2022. What is it like to be a bass? Red herrings, fish pain and the study of animal sentience. Front Vet Sci. 9:788289. doi:10.3389/fvets.2022.788289.

May RM. 2011. Science as organized skepticism. Philos Trans A Math Phys Eng Sci. 369(1956):4685–4689. doi:10.1098/rsta.2011.0177.

Mettam JJ, Oulton LJ, McCrohan CR, Sneddon LU. 2011. The efficacy of three types of analgesic drugs in reducing pain in the rainbow trout, *Oncorhynchus mykiss*. Appl Anim Behav Sci. 133(3–4):265–274. doi:10.1016/j.applanim.2011.06.009.

Michel M. 2019. Fish and microchips: on fish pain and multiple realization. Philos Stud. 176(9):2411–2428. https://link.springer.com/article/10.1007/s11098-018-1133-4.

Moccia RD, Scarfe D, Duston J, Stevens ED, et al. 2020. Code of practice for the care and handling of farmed salmonids: review of scientific research on priority issues. NFACC Scientific Committee Report. 2020 Sep. https://www.nfacc.ca/pdfs/codes/scientists-committee-reports/farmed%20salmonids_SC%20Report_2020.pdf.

Moylan L. 2022. The misanthropy of animal sentience. Academy of Ideas: Letters on Liberty, July 2022. https://academyofideas.org.uk/letters-on-liberty/.

Mulrow CD. 1994. Systematic reviews: rationale for systematic reviews. BMJ. 309(6954):597–599. doi:10.1136/bmj.309.6954.597.

Murayama O, Nakatani I, Nishita M. 1994. Induction of lateral outgrowths on the chelae of the crayfish *Procambarus clarkii* (Girard). Crust Res. 23(0):69–73. doi:10.2307/1549540.

Nagel T. 1974. What is it like to be a bat? Philos Rev. 83(4):435–450. doi:10.2307/2183914.

Naitoh Y. 1974. Bioelectric basis of behavior in Protozoa. Am Zool. 14(3):883–893. doi:10.1093/icb/14.3.883.

Nelson JS, Grande TC, Wilson MVH. 2016. Fishes of the World, 5th ed., Hoboken (NJ): John Wiley and Sons.

Newby NC, Stevens ED. 2008. The effects of the acetic acid "pain" test on feeding, swimming and respiratory responses of rainbow trout (*Oncorhynchus mykiss*). Appl Anim Behav Sci. 114(1-2):260–269. doi:10.1016/j.applanim.2007.12.006.

Newby NC, Stevens ED. 2009. The effects of the acetic acid "pain" test on feeding, swimming, and respiratory responses of rainbow trout (*Oncorhynchus mykiss*): a critique on Newby and Stevens (2008) - response. Appl Anim Behav Sci. 116(1):97–99. doi:10.1016/j.applanim.2008.07.009.

Niikawa T, Hayashi Y, Shepherd J, Sawai T. 2022. Human brain organoids and consciousness. Neuroethics. 15(1):5. doi:10.1007/s12152-022-09483-1.

O'Brien TC, Palmer R, Albarracin D. 2021. Misplaced trust: When trust in science fosters belief in pseudoscience and the benefits of critical evaluation. J Exper Soc Psych. 96:104184. doi:10.1016/j.jesp.2021.104184.

Orth DJ. 2023. Chapter 5. Pain, Sentience and Animal Welfare In: Orth DJ, editor. Fish, Fishing, and Conservation. Blacksburg: Virginia Tech Department of Fish and Wildlife Conservation. doi:10.21061/fishandconservation.

Oviedo-Garcia MA. 2021. Journal citation reports and the definition of a predatory journal: the case of the Multidisciplinary Digital Publishing Institute (MDPI). Res Eval. 30(3):405–419a. doi:10.1093/reseval/rvab020.

Passantino A, Elwood RW, Coluccio P. 2021. Why protect decapod crustaceans used as models in biomedical research and in ecotoxicology? Ethical and legislative considerations. Animals. 11(1):73. doi:10.3390/ani11010073.

Paterson BD, Spanoghe PT. 1997. Stress indicators in marine decapod crustaceans, with particular reference to the grading of western rock lobsters (*Panulirus cygnus*) during commercial handling. Mar Freshwater Res. 48(8):829–834. doi:10.1071/MF97137.

Penca J. 2022. Science, precaution and innovation for sustainable fisheries: the judgement by the Court of Justice of the EU regarding the electric pulse fishing ban. Mar Pol. 135:104864. doi:10.1016/j.marpol.2021.104864.

Poos JJ, Hintzen NT, van Rijssel JC, Rijnsdorp AD. 2020. Efficiency changes in bottom trawling for flatfish species as a result of the replacement of mechanical stimulation by electric stimulation. ICES J Mar Sci. 77(7–8):2635– 2645. doi:10.1093/icesjms/fsaa126.

Popper KR. 1963. Conjectures and refutations The growth of scientific knowledge. London: Routledge and Kegan Paul.

Pullen CE, Hayes K, O'Connor CM, Arlinghaus R, Suski CD, Midwood JD, Cooke SJ. 2017. Consequences of oral lure retention on the physiology and behaviour of adult northern pike (*Esox lucius* L.). Fish Res. 186(3):601–611. doi:10.1016/j.fishres.2016.03.026.

Puri S, Faulkes Z. 2010. Do decapod crustaceans have nociceptors for extreme pH? PLoS One. 5(4):e10244. doi:10.1371/journal.pone.0010244.

Raja SN, Carr DB, Cohen M, Finnerup NB, Flor H, Gibson S, Keefe FJ, Mogil JS, Ringkamp M, Sluka KA, et al. 2020. The Revised IASP definition of pain: concepts, challenges, and compromises. Pain. 161(9):1976–1982. doi:10.1097/j.pain.0000000000001939.

Reber AS. 2017. What if all animals are sentient? Anim Sent. 16(6):1225. doi:10.51291/2377-7478.1225.

Rehnberg BG, Schreck CB. 1987. Chemosensory detection of predators by coho salmon (*Oncorhynchus kisutch*): behavioral reaction and the physiological stress response. Can J Zool. 65(3):481–485. doi:10.1139/z87-074.

Rehnberg BG, Smith RJF, Sloley BD. 1987. The reaction of pearl dace (Pisces, Cyprinidae) to alarm substance: time-course of behavior, brain amines, and stress physiology. Can J Zool. Zool. 65(12):2916–2921. doi:10.1139/z87-442.

Rey S, Huntingford FA, Boltaña S, Vargas R, Knowles TG, Mackenzie S. 2015. Fish can show emotional fever: stress-induced hyperthermia in zebrafish. Proc R Soc B. 282(1819):20152266. doi:10.1098/rspb.2015.2266.

Rose JD, Arlinghaus R, Cooke SJ, Diggles BK, Sawynok W, Stevens ED, Wynne CDL. 2014. Can fish really feel pain? Fish Fish. 15(1):97–133. doi:10.1111/faf.12010.

Rosemberg S, Marie SK, Kliemann S. 1994. Congenital insensitivity to pain with anhydrosis: morphological and morphometric studies on the skin and peripheral nerves. Pediatr Neurol. 11(1):50–56. doi:10.1016/0887-8994(94)90091-4.

Schnell A, Browning H, Birch J. 2022. Octopus farms raise huge animal welfare concerns – and they're unsustainable too. https://theconversation.com/octopus-farms-raise-huge-animal-welfare-concerns-and-theyre-unsustainable-too-179134

Selbach C, Marchant L, Mouritsen KN. 2022. Mussel memory: Can bivalves learn to fear parasites? R Soc Open Sci. 9(1):211774. doi:10.1098/rsos.211774.

Shephard S, List CJ, Arlinghaus R. 2023. Reviving the unique potential of recreational fishers as environmental stewards of aquatic ecosystems. Fish Fish. 24(2):339–351. doi:10.1111/faf.12723.

Shields JD, Stephens FJ, Jones B. 2006. Pathogens, parasites and other symbionts. Lobsters: biology, management, aquaculture and fisheries. pp. 146–204.

Siemann LA, Parkins CJ, Smolowitz RJ. 2015. Scallops caught in the headlights: swimming escape behaviour of the Atlantic sea scallop (*Placopecten magellanicus*) reduced by artificial light. ICES J Mar Sci. 72(9):2700–2706. doi:10.1093/icesjms/fsv164.

Smaldino PE, McElreath R. 2016. The natural selection of bad science. R Soc Open Sci. 3(9):160384. doi:10.1098/rsos.160384.

Smarandache-Wellmann CR. 2016. Arthropod neurons and nervous system. Curr Biol. 26(20):R960–R965. doi:10.1016/j.cub.2016.07.063.

Smith ES, Lewin GR. 2009. Nociceptors: a phylogenetic review. J Comp Physiol A Neuroethol Sens Neural Behav Physiol. 195(12):1089–1106. doi:10.1007/s00359-009-0482-z.

Sneddon LU. 2002. Anatomical and electrophysiological analysis of the trigeminal nerve in the rainbow trout, *Oncorhynchus mykiss*. Neurosci Lett. 319(3):167–171. doi:10.1016/s0304-3940(01)02584-8.

Sneddon LU. 2003. The evidence for pain in fish. Use of morphine as an anaesthetic. Appl Anim Behav Sci. 83(2):153–162. doi:10.1016/S0168-1591(03)00113-8.

Sneddon LU. 2013. Pain perception in fish: Why critics cannot accept the scientific evidence for fish pain? (Response to Rose et al. 2012 Can fish really feel pain?). https://www.liv.ac.uk/media/livacuk/iib/fish/Response_to_ Rose_2012.pdf. - document since removed. Now available via wayback machine: https://web.archive.org/web/20150923060224/https://www.liv. ac.uk/media/livacuk/iib/fish/Response_to_Rose_2012.pdf.

Sneddon LU, Roques JA. 2023. Pain recognition in fish. Vet Clin North Am Exot Anim Pract. 26(1):1–10. doi:10.1016/j.cvex.2022.07.002.

Sneddon LU, Braithwaite VA, Gentle MJ. 2003. Do fish have nociceptors? Evidence for the evolution of a vertebrate sensory system. Proc Biol Sci. 270(1520):1115–1121. doi:10.1098/rspb.2003.2349.

Sneddon LU, Elwood RW, Adamo SA, Leach MC. 2014. Defining and assessing animal pain. Anim Behav. 97:201–212. doi:10.1016/j. anbehav.2014.09.007.

Snow PJ, Plenderleith MB, Wright LL. 1993. Quantitative study of primary sensory neurone populations of three species of elasmobranch fish. J Comp Neurol. 334(1):97–103. doi:10.1002/cne.903340108.

Stentiford GD, Bateman IJ, Hinchliffe SJ, Bass D, Hartnell R, Santos EM, Devlin MJ, Feist SW, Taylor NGH, Verner-Jeffreys DW, et al. 2020. Sustainable aquaculture through the One Health lens. Nat Food. 1(8):468–474. doi:10.1038/s43016-020-0127-5.

Stevens ED, Arlinghaus R, Browman HI, Cooke SJ, Cowx IG, Diggles BK, Key B, Rose JD, Sawynok W, Schwab A, et al. 2016. Stress is not pain. Comment on Elwood and Adams (2015) Electric shock causes physiological stress responses in shore crabs, consistent with prediction of pain. Biol Lett. 12(4):20151006. doi:10.1098/rsbl.2015.1006.

Stene A, Carrozzo Hellevik C, Fjørtoft HB, Philis G. 2022. Considering elements of natural strategies to control salmon lice infestation in marine cage culture. Aquacult Environ Interact. 14:181–188. doi:10.3354/aei00436.

Stoner AW. 2012. Assessing stress and predicting mortality in economically significant crustaceans. Rev Fish Sci. 20(3):111–135. doi:10.1080/10641262.2012.689025.

Støttrup JG, McEvoy LA. 2003. Live feeds in marine aquaculture. Oxford: Blackwell Science Ltd. 318. pgs. doi:10.1002/9780470995143.

Sutherland WJ, Spiegelhalter D, Burgman M. 2013. Policy: twenty tips for interpreting scientific claims. Nature. 503(7476):335–337. doi:10.1038/503335a.

Tigchelaar M, Leape J, Micheli F, Allison EH, Basurto X, Bennett A, Bush SR, Cao L, Cheung WWL, Crona B, et al. 2022. The vital roles of blue foods in the global food system. Glob Food Sec. 33:100637. doi:10.1016/j.gfs.2022.100637.

Tracey WD. 2017. Nociception. Curr Biol. 27(4):R129–R133. doi:10.1016/j.cub.2017.01.037.

Trevors JT. 2010. The scientific method: use it correctly. Water Air Soil Pollut. 205(S1):1–1. doi:10.1007/s11270-009-0283-6.

Troell M, Costa-Pierce B, Stead S, Cottrell RS, Brugere C, Farmery AK, Little DC, Strand Å, Pullin R, Soto D, et al. 2023. Sustainable development goals for improved human and planetary health. J World Aquaculture Soc. 54(2):251–342. doi:10.1111/jwas.12946.

Uddin SA, Rahman MM. 2015. Gonadal maturation, fecundity and hatching performance of wild caught tiger shrimp *Penaeus monodon* using unilateral eyestalk ablation in captivity. J Bangladesh Agric Univ. 13(2):315–322. doi:10.3329/jbau.v13i2.28804.

Valente C. 2022. Anaesthesia of decapod crustaceans. Vet Anim Sci. 16:100252.

Valentine MS, Van Houten J. 2022. Ion channels of cilia: *Paramecium* as a model. J Eukaryotic Microbiology. 69(5):e12884. doi:10.1111/jeu.12884.

Vera LM, de Alba G, Santos S, Szewczyk TM, Mackenzie SA, Sánchez-Vázquez FJ, Rey Planellas S. 2023. Circadian rhythm of preferred temperature in fish: behavioural thermoregulation linked to daily photocycles in zebrafish and Nile tilapia. J Therm Biol. 113:103544. doi:10.1016/j. jtherbio.2023.103544.

Verschueren B, Lenoir H, Soetaert M, Polet H. 2019. Revealing the by-catch reducing potential of pulse trawls in the brown shrimp (*Crangon crangon*) fishery. Fish Res. 211:191–203. doi:10.1016/j.fishres.2018.11.011.

Vettese T, Franks R, Jacquet J. 2020. The great fish pain debate. Iss Sci Technol Summ. 36:49–53. 2020

Walters ET. 2018a. Defining pain and painful sentience in animals. Anim Sent. 21(14):1360. doi:10.51291/2377-7478.1360.

Walters ET. 2018b. Nociceptive biology of molluscs and arthropods: evolutionary clues about functions and mechanisms potentially related to pain. Front Physiol. 9:1049. doi:10.3389/fphys.2018.01049.

Walters ET. 2022. Strong inferences about pain in invertebrates require stronger evidence. Anim Sent. 32:14. doi:10.51291/2377-7478.1731.

Weineck K, Ray AJ, Fleckenstein LJ, Medley M, Dzubuk N, Piana E, Cooper RL. 2018. Physiological changes as a measure of crustacean welfare under different standardized stunning techniques: cooling and electroshock. Animals. 8(9):158. doi:10.3390/ani8090158.

Wesołowska W, Wesołowski T. 2014. Do *Leucochloridium* sporocysts manipulate the behaviour of their snail hosts? J Zool. 292(3):151–155. doi:10.1111/jzo.12094.

Wisenden BD. 2015. Chapter 6. Chemical cues that indicate risk of predation In: Sorensen PW, Wisenden BD. Fish Pheromones and Related Cues. New York: John Wiley and Sons, p. 131–148. doi:10.1002/9781118794739. ch6.

Wuertz S, Bierbach D, Bogner M. 2023. Welfare of decapod crustaceans with special emphasis on stress physiology. Aquacult Res. 2023(1307684):1–17. doi:10.1155/2023/1307684.

Yang Y, Youyou W, Uzzi B. 2020. Estimating the deep replicability of scientific findings using human and artificial intelligence. Proc Natl Acad Sci USA. 117(20):10762– 10768. doi:10.1073/pnas.1909046117.

Yang Y, Sánchez-Tójar A, O'Dea RE, Noble DWA, Koricheva J, Jennions MD, Parker TH, Lagisz M, Nakagawa S. 2023. Publication bias impacts on effect size, statistical power, and magnitude (Type M) and sign (Type S) errors in ecology and evolutionary biology. BMC Biol. 21(1):71. doi:10.1186/ s12915-022-01485-y.

Zacarias S, Carboni S, Davie A, Little DC. 2019. Reproductive performance and offspring quality of non-ablated Pacific white shrimp (*Litopenaeus vannamei*) under intensive commercial scale conditions. Aquacult. 503:460–466. doi:10.1016/j.aquaculture.2019.01.018.

Zacarias S, Fegan D, Wangsoontorn S, Yamuen N, Limakom T, Carboni S, Davie A, Metselaar M, Little DC, Shinn AP. 2021. Increased robustness of postlarvae and juveniles from non-ablated Pacific whiteleg shrimp, *Penaeus vannamei*, broodstock post-challenged with pathogenic isolates of *Vibrio parahaemolyticus* (VpAHPND) and white spot disease (WSD). Aquacult. 532:736033. doi:10.1016/j.aquaculture.2020.736033.

Zullo L, Hochner B. 2011. A new perspective on the organization of an invertebrate brain. Commun Integr Biol. 4(1):26–29. doi:10.4161/cib.13804.

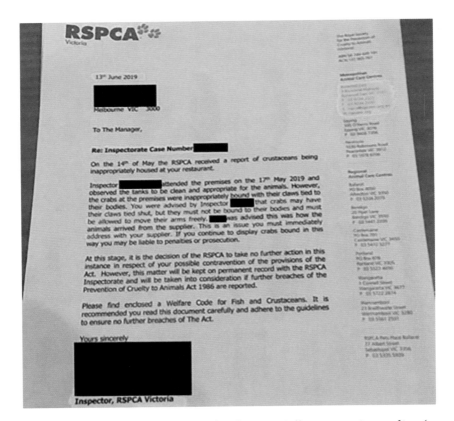

Supplement 1. This case study from Melbourne, Australia (*see Reason 8*) involved RSPCA inspectors threatening prosecution of a restaurateur holding live crabs in a display aquarium. The claw tying method used had been implemented by the live mud crab industry for many decades based not only on occupational health and safety concerns (to prevent people getting their fingers and hands crushed by crabs), but also because free claws greatly increase claw autotomy rates and allow mud crabs to injure, kill and eat other mud crabs held in the same display tanks. This example shows how great care is required to avoid 'lose-lose' situations with retrograde welfare outcomes and prosecution (even injury) of innocent people when unvalidated feelings-based welfare criteria are applied to new animal groups under suffering-centered animal welfare legislation frameworks.